Biopolitical Experience

Biopolitical Experience

Foucault, Power and Positive Critique

Claire Blencowe
University of Warwick, UK

palgrave
macmillan

First published 2012 by
PALGRAVE MACMILLAN

Palgrave Macmillan in the UK is an imprint of Macmillan Publishers Limited, registered in England, company number 785998, of Houndmills, Basingstoke, Hampshire RG21 6XS.

Palgrave Macmillan in the US is a division of St Martin's Press LLC, 175 Fifth Avenue, New York, NY 10010.

Palgrave Macmillan is the global academic imprint of the above companies and has companies and representatives throughout the world.

Palgrave® and Macmillan® are registered trademarks in the United States, the United Kingdom, Europe and other countries.

ISBN: 978–0–230–30329–4

This book is printed on paper suitable for recycling and made from fully managed and sustained forest sources. Logging, pulping and manufacturing processes are expected to conform to the environmental regulations of the country of origin.

A catalogue record for this book is available from the British Library.

Library of Congress Cataloging-in-Publication Data

Blencowe, Claire, 1981–
    Biopolitical experience : Foucault, power and positive critique /
Claire Blencowe.
        p. cm.
    Includes index.
    ISBN 978–0–230–30329–4 (alk. paper)
        1. Biopolitics. 2. Foucault, Michel, 1926–1984 – Criticism and
interpretation. I. Title.

JA80.B64 2012
320.01—dc23   2011040367

10 9 8 7 6 5 4 3 2 1
21 20 19 18 17 16 15 14 13 12

Printed and bound in the United States of America

For Carol Doran and David Blencowe,
with thanks for everything

Contents

Acknowledgements

A great many people, places and gatherings have contributed to the making of this book – which started life as a PhD thesis at the University of Bristol (funded by the ESRC). Throughout this time Thomas Osborne was a supportive, insightful and inspiring supervisor. Whilst Tom would refuse as a matter of the highest ethical concern to become anyone's 'mentor', he has shown me an enormous generosity of humour, time and trust, giving me not simply the freedom, but the *courage* to pursue my independent lines of thought, whilst always being on hand to offer direction or a pep talk, to test my reasoning or to beat my work into comprehensibility. My PhD examiners, Marianne Fraser and Gregor McLennan, also contributed a great deal in the way of critical insight, shaping, editing and encouragement. I am very grateful. The *Authority Research Network* has been crucial source of inspiration, friendship, energy and critique contributing to the final formation of the ideas in this book: my thanks to Aécio Amaral, Julian Brigstocke, Leila Dawney, Naomi Millner, Tehseen Noorani and Samuel Kirwan – long may our collaborations continue! I am also very grateful to Vivienne Jackson, Nasar Meer, Katherine Smith and anonymous reviewers from *Theory, Culture & Society* and *History of the Human Sciences* for their comments on chapters. More people than I can mention have given me invaluable support whilst working towards the completion of this text: a special thanks to all those Bristolians who participated in the *Critical Theory Group, Chow Tours, Lazlo Kiss* or *Research Club*; to Ranji Devadason, Asahi Takano, Ruby Narbough and Elinor Stiles for helping me through various hardships; to my former colleagues and Culture Power & Biopolitics students at Newcastle University; and to my immensely supportive and patient family, Carol and Danny Doran, David and Lynda Blencowe, and Helen, Simon and Mae Blencowe-Fraser. Above all, however, I would like to thank my in-house philosopher, editor, cheer-leader and best friend Julian Brigstocke, with gratitude and love.

Chapter 3 of this book was originally published as 'Foucault's and Arendt's Insider View of Biopolitics a Critique of Agamben' in *The History of the Human Sciences*, 2010 and is reproduced here with the permission of Sage. A truncated version of Chapter 5 originally appeared as 'Biology Contingency and the Problem of Racism in Feminist Discourse' in *Theory Culture & Society*, 2011 and is also reproduced here with the permission of Sage.

Introduction

This book offers an expansive, materialist, interpretation of Michel Foucault's critical account of biopolitical modernity. In the late 1970s Foucault claimed that society's 'threshold of modernity' had been reached at the moment when *life* entered political strategy and history, with the emergence of biopolitics and the 'incorporation' (in-corporealisation) of power (1978:143; 2000a:125). Over the past decade a conception of modernity as biopolitical, which draws upon Foucault's work, has become increasingly significant for sociological and political theory. This book puts forward an original interpretation of what Foucault's work on biopolitics is about, contending that biopolitics should be understood as a historically specific formulation of *experience* and *embodiment* – a formulation or 'framing' that constitutes life as an immanent ground of meaning, truth games, ethics and political reason. A focus upon experience and positivity produces a theory that is better able to explain the 'hold' or 'viscosity', the productiveness and the appeal, of biopolitical discourse. I argue that Foucault's critical vision is one that illuminates the positivity, appeal and productiveness of biopolitics – not in order to *celebrate* that positivity, but to bolster our freedom with respect to its charms.

For Foucault the history of biopolitics is a history of bodies: how they have been invested and connected with each other, with the world, with the transcendent, the transcendental and the immanent. It is the history of a formation of embodiment and experience that focuses questions of value – questions of epistemology, political reason and ethics – upon the manifestation, protection and perpetuation of life. Biopolitical, biological life is immensely productive, connective and processual; it is immensely vital, and its emergence as a domain of reality gave rise to a broadly vitalist array of rationalities, motivations and ethics.[1] This

1

means that biopolitics pertains to something more general than 'the politics of the body', invoking, instead, something like a 'horizon of culture', a 'field of visibilities', a 'game of truth' and a 'cartography of force and affect'. Biopolitics is, however, far less totalising than some proclaim. It is not a type, stage or hidden truth of Society, the West or modernity, nor is it a 'historical epoch'. There are numerous rationalities at play in modern societies, none of which takes hold of society as a totality. Indeed biopolitical rationalities and economies of experience are themselves diverse and malleable, incorporating *a* multiplicity not *the* totality of modern political institutions, rationalities and ethics.

The study is motivated by a concern with a number of issues pertaining to contemporary sociality. These motivations include a drive to develop a historical and pluralistic perspective on the ethoses of vitalism, creativity and new materialism that are giving shape to social and political theory 'after post-structuralism'; a related desire to problematise and complicate an ethic of progressivism that celebrates contingency and creativity for their own sake, failing to differentiate between different images and productions of differentiation, life or creativity; and a determination to contribute to the exploration of the embodiment of present political discourse, exploring what present formations of materiality and experience mean for the parameters and possibilities of empowerment, collective politics and domination and asking what the history of biopolitical discourse can tell us about contemporary biopolitics. However, whilst these issues motivate the study they are *not* the object of the study. The book is a theoretical and genealogical exploration of biopolitical discourse and experience putting forward expansive interpretations of Foucault's writings on biopolitics and related texts. It is emphatically *not* an empirical investigation of contemporary biopolitics or an attempt to *apply* the theory of biopolitics to contemporary phenomena. Reflections on the implications of my analysis for an understanding of contemporary biopolitics or sociality remain, as such, highly speculative.

Nonetheless a commitment to sociality runs throughout this study. I adopt a specifically 'sociological' perspective with respect to biopolitics. This is neither to say that I adhere to some particularly sociological brand of empiricism, nor to say that I hold to a predetermined set of categorical priorities (class, gender and race, or social facts). Rather, I draw on a specifically sociological perspective and tradition, or ethos, insofar as I place an ontological, normative and explanatory priority on *processes of sociality*: trans-individuality, creating connections, influencing and interacting. The book assumes not only that discourses, technologies and objects of knowledge are produced through sociality but also that making and manifesting connectivity, sociality and influence constitute the drive, allure and substance, or 'embodiment', of experience, affect and subjectivity. From this perspective the emergence of biological knowledge appears, above all, as a refiguration and invention of relationships, of influence and of connection.

Biopolitics and 'life' as modern experience

The originality of my interpretation of Foucault's account of biopolitics rests in conceiving of biopolitical modernity in terms of a *historical formation of experience*. This conflicts with interpretations of biopolitics that claim it represents a total or ahistorical truth of capitalist or Western societies, as well as with those that tie biopolitics too closely to the history of biological science. Foucault's account of biopolitics addresses the historical formation of knowledge – of biological knowledge. However, knowledge, for Foucault, is a very general category of organisation; organising experience, space and time, truth games and relationality, as well as science and expertise. As Gilles Deleuze puts it, knowledge, for Foucault, 'is not science and cannot be separated from the various thresholds in which it is caught up, including even the experience of perception, the values of the imagination, the prevailing ideas or commonly held beliefs' (1988:44). Certainly the history of biopolitics is, as Nikolas Rose suggests, the history of modern scientific expertise, its utilisation in governance and in the development of a politics of the somatic. Certainly the history of biopolitics does, as Giorgio Agamben argues, describe the context wherein European States developed concentration camps and orchestrated thanato-political eugenics and ethnic genocides. But the history of biopolitics is *also* the history of the positivity – the experiential, embodied, perceptive and affective forces – of modernity and the formations of value and evaluation that these entail. It is the history of a proliferation of experience, of contingency and creativity; a politicisation of embodiment and life; and a production of life – vitality, creativity, health, reproduction, security and evolution – as *values,* inspiring and motivating ethical, epistemological and political work. Through a focus on the experiential dimensions of biopolitics, my interpretation illustrates significant and largely overlooked *continuities* of concern across Foucault's oeuvre. The concern with 'life' as a historically produced category and with the role of limits in the constitution of life and experience stretches across Foucault's oeuvre, from *The Birth of the Clinic* to *The Care of the Self.*

I will elaborate upon the concept of 'experience' below. For now suffice it to say that experience is, for Foucault, about truth games, originations of meaning, embodiment, investments and aesthetics, which are historically constituted in organising knowledge (fields of visibility, grids of intelligibility, language, epistemology). Experience, for Foucault, is processes that constitute, transform and transfigure what counts as a subject and what counts as an object (Jay, 2006:400; Foucault, 2000c:257). It is what is seeable, what is sayable, and what can affect, as well as what can see, what can speak truth, and what can be affected (as both objects and subjects are constituted in the process of experience). Although Foucault strongly rejects romanticist and phenomenological valorisations of experience and authenticity (with their ahistoricism and humanism) he does describe his own motivation in terms of the pursuit of experience. With this in mind he refers to experience as the transfiguration of subjects and objects, trying to reach the limits of the liveable, and as 'a project of desubjectivation' (Foucault, 2000c:241–3). The emergence of modern biology is a fundamental event in the history of experience because modern biology 'places man as a living being in question' (Ibid.:256). The movement beyond the subject, desubjectivation, is internalised in biological life, creating an immanent, 'trans-finite', dimension. With the emergence of modern biology, life escapes from the general laws of being, becoming the nucleus of both being and non-being, and 'the experience of life is thus posited as the most general law of beings', constituting an 'untamed ontology' (Foucault, 1970:278). Experience, it seems, not only transforms, but gains a radical new significance with the emergence of modern biological knowledge.

As will be demonstrated in Chapter 1, Foucault argues that life, as conceived with modern biology and biopolitical thought, did not exist prior to the nineteenth century. The terms 'biology' and 'bio', in Foucault, do not then refer to 'the opposite of culture' or to individuatable somatic bodies or to any ahistorical category. Rather 'biology' refers to a particular knowledge that developed in the nineteenth century, whilst 'life' (and the 'bio' of 'biopolitics') refers to 'trans-organic' serial phenomena that take place at the level of the population. Life escapes the laws of being; it comprises an immanent domain of transcendence, a plane of emotional, epistemic and economic investment; it is 'a quasi-transcendental' (Foucault, 1970:250–1). It does, as such, produce a dimensionality within the present: an epistemic, affective and perspectival depth within the finite passing world.

As will be argued in Chapter 2, the depth that life creates consists of relationships of connection, responsibility and creativity. These new relationships and capacities are deployed in nineteenth-century theories of reproduction, evolution and degeneracy. The biological formation of embodiment entails a 'biological responsibilisation' of bodies and practices, such that the most intimate of activities can become of enormous public concern. It is, as such, immensely *politicising*, radically multiplying demands, and possibilities, for political intervention, as well as capacities, contingency and experience. The history of biopolitics is not just about the development of certain political institutions or the political uses and abuses of biological science. It is also about a historical reconstitution of experience, of the organisation of meaning, of what matters, of authority and games of truth, such that *immanent* processes, embodiments and manifestations of vitality or intensity become salient in the formation of perception and judgement. As Foucault would sometimes put it, they become salient in the formation of 'subjectivity'. Biopolitics, in Foucault's work, is about 'modernity' and the experiential context of modernism, modernist ethics and aesthetics, as well as of modern science and political institutions. As Chapter 3 will show, Foucault and Hannah Arendt concur in associating biological politics with a general valorisation of mortal life and the experience of process, as well as with a kind of immanentisation of the eternal life that was the object of Christian theology and pastoral power. As Chapter 4 will suggest, Foucault's theories of biopolitics are relevant in the analysis of a range of political discourses that are not generally thought of as 'biological'. Biopolitical economies of experience may, for example, be operative in the socially constructivist discourses of neo-liberalism, cultural racism and, as will be suggested in Chapter 5, feminism. Indeed Chapter 5 sets out a specific example of modern biopolitical discourse (in the shape of twentieth- century feminism), illuminating the positivity and appeal of biopolitical experience, as well as demonstrating the value of Foucault's positive critique of biopolitics contra the example of negating materialist-feminist ideology critique.

Positive critique

A central claim of the book is that Foucault's exploration of the biopolitical nature of modernity should be understood as an exercise in 'positive critique'. 'Positive critique' encapsulates both a pluralizing approach to history and an experiential, embodied, pragmatic (in the Deweyian sense) approach to political ideas, investments and discourse. Foucault develops a positive critique of biological knowledge and politics that stands in contrast to theoretical tactics such as 'ideology critique' that are rooted in the negation of biological knowledge.

'Positive critique' is related to the conviction that experience – at least in the context of modernity – is a matter of the processuality, connectedness and openness of relationships and forces in the world, rather than the embedding, continuity, stability or security of a subject. I deploy this term in an attempt to capture a critical attitude that is a crucial, but often overlooked, aspect of genealogical work upon modernity. Positive critique is 'positive' because it interprets events in terms of the expansion of forces, capacities and experiences of empowerment. It is interested less in demonstrating the badness or fallacy of a given political discourse or institution (in the work of negative critique) than it is in illuminating the 'hold' that political discourses and institutions have upon people: what makes power acceptable; its appeal or allure. This appeal and allure is always understood in terms of productivity, the production of things, pleasures, affects and capacities, especially knowledge. Positive critique is also 'positive' because it is only concerned with productivity, addition and pluralization. This is not only the case in the analysis of political situations but also in the description of history. Positive critique treats history as a series of additions in which new events, new rationalities, new economies, interact with and transform those that existed already. It does not paint a linear picture of history wherein new things are assumed to displace, resolve or substitute for the old.

'Positive critique' describes an aspect of genealogical enquiry. As David Owen argues, genealogy is about freeing oneself and others from captivity to a given perspective, illustrating that it is but one perspective, that it is historical, that it could be otherwise, and thus enhancing critical freedom (Owen, 2002). Genealogies also describe history in terms of bodies, forces, struggle and strategies – not only describing knowledge as historical but insisting that it is a history constituted in embodied struggle, strategy, combat and resistance. I am interested in the link between these two elements that are addressed by genealogy: the history of knowledge and the dynamics of forces and embodiment. That link pertains to understanding the power of knowledge in terms of an appeal, or stickiness, that results from the augmentation and investment of forces and capacities, the experience (if not exactly the reality) of empowerment, transformation or expansion. Genealogy does not only demonstrate that perspectives are always historical. By illuminating what has been at stake in the struggles by which *this particular* perspective did in fact take hold, genealogy can also help to ascertain why this perspective is appealing or forceful in the present. It is a 'positive critique' both because it attempts to grasp a given discourse in its own positivity (rather than subordinating it to the assessment of an alternative truth game) and because it assumes that the history of knowledge will be the combative history of 'positive' productive or expansive forces.

Positive critique is *not* positive in the sense of being happy or in wishing to celebrate positivity. Positive critique is about the positivity of the world but this does not mean that it is 'positive about' the world as given – it is a *critical* strategy. Although positive critique might deploy a kind of minimally vitalist ontology, it is not of that genre of vitalist methodology which is concerned to capture and celebrate the vitality and liveliness of the world for the sake of that liveliness (e.g. Thrift, 2001). Positive critique is 'critique' because it is a work upon the limits of the present, exposing economies of experience that give power its power. It describes knowledges, discourses and institutions in terms of their positivity and actual intelligibility rather than denouncing them on the grounds of their destructiveness, irrationality or fallacy. However, it does so in order to *detach* us from these mechanisms, to open up and pluralize our perspectives, to generate space and demand for politics and critical judgement (Foucault, 2000c:244–5). Positive critique is 'denaturalising', 'defetishising' and 'detaching' without being condemnatory, Othering, dichotomising or dismissive. If it is motivated by a normative vitalism, then that is with respect to *itself* as work upon the limits of the present, vitalising that present. Positive critique can be understood as normatively vitalist insofar as it is itself a work towards opening the world, loosening perspectives, generating the creative activity of autonomous judgement and political combat. As Foucault might have put it, positive critique 'produces more opportunities to error', which in the terms of Georges Canguilhem is to say that it 'increases life' (see Foucault, 2000b).

The positivity of biopolitics

A genealogical investigation of political discourse is not, then, simply the demonstration that such discourse is historically constructed. It is also an attempt to capture the *positivity* of that discourse. One of the great contributions of Foucault's work for contemporary social science is that it illuminates, and helps us to get to grips with, the positivity of biologistic thought (a mode of reasoning that has, rightly, been a central concern for critical sociology for many decades). Foucault's work on modern knowledge and political discourse suggests that the biological is appealing not because it is conservative, primordalist or preformist but because of the close agreement between biology and the proliferation of contingency, progressivism, transformation, modernism and certain experiences of empowerment.

A widely held view of biological politics and social ontology assumes that it is, effectively, 'preformist', describing the world and its biological differences as fixed in a preordained natural order. This relates to the notion that biologism has been successful as a political discourse in the past because it had fed a desire for epistemic security and conservative experience. That view of biologism remains important for contemporary sociology because it forms a common point of oppositional reference in explanation of the significance and political implications of socially constructivist ontologies – inviting a valorisation of contingency, processuality and indeterminacy. Foucault's presentation of biopolitics as a positive, productive, processual discourse and rationality challenges the association between biologism and deterministic conservatism. This issue runs throughout the book but comes to a head in Chapters 4 and 5, where I reconsider the historical relationship between contingency, social constructivism and biologism. I tentatively suggest that the characterisation of biological politics as conservative and determinist has served not to distance social ontology from key assumptions of biopolitics so much as to mask *continuities between* biologism and social constructivism.

Having set out the problematic of this book in broad outline we will now move on to consider the context of that problematic. The problematic of the book is the critique of biopolitical modernity in terms of the formation and economy of experience. In the following section I will attempt to illuminate what is at stake in the critique of modernity as biopolitical through a discussion of its relationship to Marxist ideology critique and Weberian critique of modernity-as-rationalisation. I will then address the concept of experience in Foucault's work.

Biopolitics and the critique of modernity

This book is interested in the theory of biopolitics in the *critique of modernity*, loosening the grip of present perspectives and modes of investment – present power/knowledge. The genealogy of critical theories of biopolitics stretches into a number of critical traditions in philosophy and social and political theory. We could, for example, look to Kant or to Nietzsche for a suitable starting point from which to map a trajectory of critique that ends with the theory of biopolitics. From the perspective of this book, which adopts a sociological ethos, Marxist ideology critique and Weberian critiques of modernity as rationalisation are particularly salient. Foucault has explicitly situated his critique of modernity and his economy of power in contrast to these two traditions (Foucault, 2000b; 2000c; 2000e). There is clearly a considerable alignment or resonance between these two older critical approaches and the theory of biopolitics. The latter would best be seen as an addition to, and pluralization of, Marxist and Weberian critiques of modernity. The biggest issue that Foucault takes with both the Marxist and Weberian critical projects pertains to their *singularising* approach (to power and history, or to knowledge, experience and rationality).

Marxism and ideology critique

Theories of biopolitics have grown in part out of a disillusionment with Marxism as a totalising critique of modernity, power and domination, and especially with 'ideology critique' as a strategy for attaining political transformation through the pursuit of knowledge. Foucault claimed that he was attempting to develop a new economy of power focused upon bodies, knowledge and subject formation. He often explained this project by way of a problematisation of the form of Marxism that was dominant amongst European left wing intellectuals throughout the first decades of his career (Foucault, 2003a; 2000c; 2000h). Foucault's problematisation of Marxism is tied to the experience of daily struggles to transform power, which makes the concrete nature of power visible, and of situations of domination that have relatively little economic significance – to psychiatric internment, mental normalisation and penal institutions (Foucault, 2000a:117). It is, as such, tied to the historical moment of '1968' which has come to represent the emergence of a new generation of political struggle that contested the privilege of class in the interpretation of domination and brought the diverse, dispersed, often symbolic or embodied and fundamentally ambiguous nature of power to the fore.

As will be argued in Chapter 3, Arendt similarly positioned her work against ideology critique and the overemphasis upon the truth or falsity of political ideologies. Referring to Nazi ideology she noted:

> if a patent forgery like the 'Protocols of the Elders of Zion' is believed by so many people that it can become the text of a whole political movement, the task of the historian is no longer to discover a forgery. Certainly it is not to invent explanations which dismiss the chief political and historical fact of the matter: that the forgery is being believed. This fact is more important than the (historically speaking, secondary) circumstance that it is a forgery. (1968:7)

Arendt continued to use the term 'ideology' but she redefined it, limiting the term strictly to totalising political discourses which, she claimed, are characterised by a relationship to temporality or historicity, not to truth. 'Ideologies', she writes, 'are never interested in the miracle of being. They are historical, concerned with becoming and perishing, with the rise and fall of cultures, even if they try to explain history by some "law of nature"' (1968:469). Arendt developed her account of modern politics as the rise and emancipation of biology (or 'labour') in the context of an explicit – although avowedly sympathetic – critique of Marx (Arendt, 1998:79–135). She argued that Marx failed to differentiate between alternative types of the *vita activa* – between different types of active life. He confused the drudgery of labour (biological life) with the artistry of work (world building or craft) and the autonomous creativity of political action (speaking or acting in a public of autonomous actors) (Ibid.:306) – a confusion which had dire consequences for political Marxism in Arendt's view. Marxist politics do not challenge, can even extend, the destructive, oppressive, normalising dominance of biological labour in modern life, as Arendt sees it.

Foucault referred to a number of key differences between Marxist analysis, especially ideology critique, and the 'new economy of power' that he was developing, of which the theory of modernity as biopolitical was a major part. Like most critics of Marxism, Foucault stresses the point that there are many forms of domination that are insignificant from the perspective of economics or class struggle but are, nonetheless, immensely important. Most of his comments, however, refer to the Marxist conception of *knowledge* and thus to the notion of ideology. In Foucault's view the notion of ideology is premised upon a failure to understand the historicity of truth and the primacy of knowledge in the formation of power relations. He describes the Marxist notion of ideology thus:

> In traditional Marxist analysis, ideology is a sort of negative element through which the fact is conveyed that the subject's relation to truth, or simply the knowledge relation, is clouded, obscured, violated by conditions of existence, social relations, or the political forms imposed on the subject of knowledge from the outside. Ideology is the mark, the stigma of these political or economic conditions of existence on a subject of knowledge who rightfully should be open to the truth. (Foucault 2000d)

To be clear, it is not my intention to imply that ideology critique is the sum total of Marxist analysis and critique of modernity, but simply that ideology critique is particularly problematic from the perspective of positive critique and the Foucauldian theory of biopolitics. In many respects Foucault's theory of biopolitics should be seen more as a development from, than a rejection of, Marxist critique. The theory of biopolitics is, at the least, an expansion upon and pluralization of critical political economy.

In contrast to ideology critique, Foucault contends that there are a *plurality* of rationalities, economies and forms of power at play in the productions of modernity. In particular, he stresses that knowledge is not reducible to economic power, comprising its own, equally important, economy. 'If the accumulation of capital was one of the fundamental traits of our society' he argues 'the same is true of the accumulation of knowledge' (2000c:291). Moreover, he argues (against the notion of ideology) that power cannot be contrasted with or separated from truth, as though power were the corruption of truth. Truth, Foucault argues, is produced though power and *is itself* power. '[T]ruth isn't the reward of free spirits, the child of protracted solitude, nor the privilege of those who have succeeded in liberating themselves. Truth is a thing of this world: it is produced only by virtue of multiple forms of constraint' (2000a:131). There is no subject that could know, or could experience, outside of the historical formation of truth games and subjectivity/experience. Subjects are perpetually produced, along with objects, in the acts and organisation of seeing, speaking and knowing. Knowledge is the historical condition of subjectivity, experience and truth. Truth should be considered a part of the complex, historically constituted, constituting, economy of power and productive force:

> The problem [for intellectuals] is not that of 'changing people's consciousness – or what's in their heads – but the political economic, institutional regime of the production of truth. It's not a matter of emancipating truth from every system of power (which would be a chimera, for truth is already power) but of detaching the power of truth from the forms of hegemony, social, economic, and cultural, within which it operates in the present time. (2000a:133)

To critique modernity from the perspective of Foucault's 'new economy of power' is to pay attention to historical formations of knowledge which produce truth games, ways of being a subject, ways of experiencing – it is not to 'speak the truth' to or against a corrupting, falsifying power. Moreover it is to pay attention to a diverse and distributed network of power in its concrete, creative formulations, not to work out how every noted situation of power can be explained by a common cause. The theory of biopolitics emphasises the role of knowledge – specifically biological knowledge – in the formation of modernity and its relations and productions of power. It is interested less in the exploitation of bodies than it is in the production of embodiment, and less in the corruption of the autonomous subject than in the production of subjectivity and autonomy. A number of crucial differences emerge between the Foucauldian-biopolitical and the Marxist critiques of modern power.

Most notably, the critique of modernity in terms of biopolitics places a series of issues at the heart of political economy that are relatively insignificant from the perspective of capital flows and accumulation. Sexuality, race, emotion, health, security and autonomy appear at the centre of biopolitical economy – not as *means* of exploitation but as *ends* of power. This focus comes in tandem with a positive conception of power, a conception of power as 'a productive network that runs throughout the whole social body', not 'a negative instance that operates through repression' (Foucault, 2000a:120). The theory of biopolitics describes the 'hold' of modern power and knowledge in terms of the ways in which it *expands* and enables peoples' bodies, creates pleasures and generates capacities. This stands in contrast to ideology critique, which, in seeking to demonstrate the fallacy or irrationality of beliefs and relations, can be dismissive of people's real investments in power and of real subjectivity (which is constituted within, not outside of, power and history).

An overlooked aspect of Foucault's account of biopolitics is that he *does* place considerable emphasis upon the role of the bourgeoisie, and bourgeois class hegemony, in the formation of modern power – another topic that will be explored in Chapter 2 of this book. Whereas Marxist analysis is interested in bourgeois hegemony as a relationship of *exploitation*, Foucault is interested in *a new productivity of power* that was developed by the bourgeoisie, primarily (in the first instance) for application to the bourgeois class *itself*. A bourgeois power developed not as a mechanism of deduction but as an intensification and affirmation of the bourgeoisie's own life, health and body. Whilst bourgeois hegemony was central to the formation of biopolitics, according to Foucault, this does not mean that biopolitics can be understood as a situation of class enmity. Bourgeois power, according to Foucault, is not 'the exploitation of the proletariat' or 'the ownership of means of production'. It is an economising, intensifying, vitalising array of productive mechanisms, procedures and capacity. 'What was formed', he claims, 'was a political ordering of life, not through an enslavement of others, but through an affirmation of self' (Foucault, 1978:123). This power, which first developed in the eighteenth century, was, Foucault claims, extended to the whole of society (in European nation states) by the end of the nineteenth century. I will argue in Chapter 2 that we can see nationalism as a generalisation of this self-affirming, self-intensifying, life-maximising, bourgeois power. Certainly Foucault claims that biopolitical power was extended to the whole of society at the time of the establishment of the nation-state and that this remains a productive, not deductive, power. After the generalisation of biopolitical power to the whole population Foucault suggests that the hierarchical relationship between classes was produced and maintained through the production of *biological variations*, such as differentiations in degrees of sexual repression and degeneracy (Ibid.:129). Biopolitical class relations are characterised by biological differentiation, connection and regulation, rather than by class enmity or dialectical struggle. In effect Foucault is suggesting that modern classes were produced as *races*; as variations within, or fragments of, a biological population.

Racism, sexuality, health and the production of autonomy are essential to the biopolitical economy of nation states. Foucault argues that racism is 'inscribed as the basic mechanism of power, as it is exercised in modern States' (2003b:254). On the one hand nationalism appears as a kind of racist biological self-affirmation. On the other, a racism that is *internal* to national populations acts as a sort of 'converter' enabling negating thanatopolitical practices to be experienced, or performed, as productive maximisations of life (Ibid.). The self-affirming, intensifying biological population and its racisms are produced in significant part through the historical formation of sexuality, alongside a plethora of normalising and regularising discourses. The political economy of biopolitics is not interested in the ways in which power denies people's freedom, but rather in the ways in which formulations of emancipation and empowerment are *produced through power*, binding people to power and to productions (and regularisations) of collective embodiment. In Chapter 5 I will argue that feminism is a biopolitical discourse, emphasising the affinity between biopolitical power and the production of emancipation (and racism). The theory of biopolitics is, then, considerably more attuned to the *diversity* of power relations that are in play in modern societies than is traditional Marxist analysis. It places health, race, sexuality, security and autonomy at the centre of political economy and critique.

The critique of modernity in terms of biopolitics can be seen as a pluralization of the concerns of Marxist critique – pluralizing the domains of reference of political economy, contesting the singularity of the significance of capital accumulation, of rationality, of the subject of knowledge, and of history. Biopolitics is neither a historical substitution for capitalism nor a more accurate interpretation of what capitalism is about. Rather, modernity is both capitalist *and* biopolitical, and many other things besides. Whilst the theory implies a considerable rejection of Marxist historical ontology and politics of knowledge, it adds to, rather than attempting to displace, Marxist political economy. Like Marxist critique, the theory of biopolitics is concerned with an analytics of production, accumulation and force.

The Weberian critique of rationalisation

Arguably, Foucault's account of biopolitical modernity shares a path away from Marxist analysis with Max Weber and a range of critical sociologists who have found inspiration in his conception of modernity as a sometimes irrational process of rationalisation. Like Foucault, Weber draws inspiration from Nietzsche as well as Marx. Like Foucault, Weber is interested in the history of modern rationality, the role of values and knowledge in the formation of political community (authority), and the ambiguous – autonomising/imprisoning – effect of processes of 'rationalisation' (Gordon, 1987). Foucault describes as 'Weberians' 'those who set out to trade off [*relayer*] the Marxist analysis of the contradictions of capital for that of the irrational rationality of capitalist society' (2000e:229). To this we should add the idea that rationality somehow dominates, destroys or denies life, experience and values in the context of capitalist modernity – that instrumental-reason dominates and destroys value-reason. Zygmunt Bauman's conception of modernity as a perpetually renewed, inevitably failing, attempt to eliminate ambiguity (Bauman, 1989), and Jürgen Habermas's critique of the 'colonisation of the life world' (Habermas, 1987), are examples of a broadly Weberian approach to the critique of modernity. Both resonate, in a limited sense, with Foucault's work on the problems of modernity – especially with his description of a 'disciplinary power' that invested bodies with normalising and economising power/knowledge from the seventeenth century onward (Foucault, 1977).

Whilst Foucault is keen to stress that he is not a 'Weberian' (2000e:229–31), he does express an affinity with one of the great schools of Weberian thought: the Frankfurt School. Foucault states that the Frankfurt School had been the first to raise a number of problems that were also his, or 'our', problems: especially the problem of the effects of power in their relation to modern Western rationality. In setting out that problem Foucault states that this rationality must be considered necessary to the economic and cultural results that characterise 'the West' and that this rationality seems inseparable from 'the mechanisms, procedures, techniques, and effects of power that accompany it and for which we express our distaste by describing them as the typical form of oppression in capitalist societies' (2000c:273). He continues:

> Couldn't it be concluded that the Enlightenment's promise of attaining freedom through the exercise of reason has been turned upside down, resulting in a domination by reason itself, which increasingly usurps the place of freedom? This is a fundamental problem we're all struggling with...And, as we know, this problem was isolated, pointed out by Horkheimer before all the others. (2000c:273–4)

Foucault does, then, share in a broadly Weberian project of critique concerning the role of modern rationality in the production of modern power, domination, objectification and un-freedom.

Foucault is not, however, a Weberian (and nor is he in general agreement with Frankfurt School members), because he rejects the singularity of the concept 'rationality' and 'rationalisation', preferring instead to talk about multiple histories of rationalities and reasons, defined instrumentally and relatively (something is rational, reasonable or subject to games of truth, *with respect to* this or that regime of knowledge). In an interview he stated:

> I don't think I am a Weberian, since my basic preoccupation isn't rationality considered as an anthropological invariant. I don't believe one can speak of intrinsic notion of 'rationalisation' without, on the one hand, positing an absolute value inherent in reason, and, on the other, taking the risk of applying the term empirically in a completely arbitrary way. I think one must restrict one's use of this word to an instrumental and relative meaning. (2000e:229)

The accuracy of Foucault's interpretation of the Weberian conception of rationality as an anthropological invariant is debatable. However, the difference that he is alluding to here certainly does distinguish the biopolitical critique of modernity from that of representatives of the first generation Frankfurt School, Habermas and Bauman, insofar as the latter treat 'modern rationalisation' (and indeed modernity) as a *singular* event or type of process.

(At least) two different modern rationalities

Weberian critique understands modern 'rationalisation' in terms of a singular conception of rationality. Modern instrumental-reason might be opposed to value-reason in Weber but modern rationalisation is about the development and encroachment of instrumental reason alone. In Foucault's account of modernity at least two different forms of rationality and rationalisation appear. These are separated by the historical moment of their emergence, the regimes of knowledge by which they are organised, their frame of reference, their sites of veridification (locations in which truth and falsity are conferrable), and the formations of embodiment that they apprehend. In addition to a classical modern rationality that is directed at the imposition of an artificial, rational order upon man's nature, a biopolitical rationality is directed to the generation, manifestation, protection of, and deferral to, life, vitality, creativity and spontaneity.

The Weberian account of modern rationalisation as the encroachment of instrumental reason and the image of modernisation as a (necessarily doomed) attempt to establish a common, disambiguous order which extends the clarity of reason throughout social life is paralleled in Foucault's account of 'classical reason' and the police state, which emerged in sixteenth and seventeenth century Europe alongside the normalising disciplinary institutions and practices that have proliferated throughout European societies, their (former) colonies and beyond since that time. This classical, classificatory, policing reason shares a regime of knowledge with natural history, not with modern biology. Natural history is not interested in the historicity of nature – in fragmentation, evolution or destruction (in 'animal' nature) – but in the extensive ordering of superficial difference and similarity, in a 'botanical' nature calmly mapped out in two dimensional space (Foucault, 1970). The police state attempts to economise its territory and people by controlling processes, imposing rationality, creating order (Foucault, 2007). Whilst Foucault does not develop the argument it would be easy to draw parallels between this disciplinary, rationalist modernity and Calvinism, to which Weber gives such a prominent role.

The conception of modernity as biopolitical suggests that there is an additional modernity, an additional, distinct, modern rationality, an alternative 'programme'. The threshold of this modernity was reached in the nineteenth century, when man began to consider himself as a living species. This modernity is affective, processual, immanentist, intensive, vitalist – it is (in the first instance) bourgeois and it is nationalist.

Biopolitical rationality shares a regime of knowledge with modern biology, political economy and linguistics. It is a rationality that is concerned with processes, forces of expansion and transformation, which are self-generating, self-perpetuating and which are meaningful, or at least logical and normative, in and of themselves. These processes do not need to have an artificial order imposed upon them for them to be sensical and economical, they are self-generating order – processes such as biological evolution. The surface of things is relatively insignificant from the point of view of biopolitical reason because the important processes, such as the unfolding of life, take place in an obscure depth of historic time. Biopolitical modernity means: the emergence of liberalism, a political strategy that aims to regulate already existent, self-regulating, economic and public processes; the emergence of nationalism, the self-affirmation and self-vitalisation of the Nation; and it means the emergence of eugenics, theories of degeneracy and new variants of racism informed by evolutionary theory and contributing to a dynamic view of history. Although there has not been the opportunity to develop such an analysis in this book, we *could* associate this additional biopolitical modernity with the history of Protestantism as well. To do so we would not look to the rationalism and individualism of Calvin, but to the Great Awakenings and evangelical revivalism of the eighteenth century – the emotive, expansive, welfareist turn that gave rise to the Methodists, the Baptists and the Quakers, movements that created a direct emotional relationship with God, the Church and the world, which were radically expansive, carrying the gospel and the new Protestant message to the furthest reaches of the European Empires, and which were decentralised, with the power to carry the creed extended to laymen, who acted as both lay preachers and missionaries. Biopolitical rationality addresses a world comprised of expansive, affective, embodied processes. It is not an economising rationalisation of bodies, rendering them docile, objectified, disciplined. Rather it is a rationality of trans-organic, vital processes and force that must be, to an extent, respected (*laisser faire*), and that can be harnessed and managed but not artificially created. It is subjectifying, not objectifying. Chapter 1, which describes the specificity of the 'bio' in Foucault's theory of biopolitics by way of a comparison between modern biology and natural history, and discipline and biopolitics, will elaborate on these distinctions between disciplinary and biopolitical modernity.

Rationality, experience, freedom

The distinction that Foucault draws between the Weberian approach and his own also pertains to a series of dualisms or dichotomies that

Foucault wants to do away with, including the opposition between freedom and power, between abstract/objectifying rationalism and subjective experience, and between nature and history. All of these oppositions depend upon a certain metaphysical, or at least ahistorical, conception of the subject. Foucault denies that there is an opposition between, on the one hand, the intellect, or intellectualisation, and on the other life, freedom or experience. 'The fact that man lives in a conceptually structured environment', he writes, 'does not prove that he has turned away from life, or that a historical drama has separated him from it – just that he lives in a certain way, that he has a relationship with his environment such that he has no set point of view toward it, that he is mobile' (Foucault, 2000b:475).

Foucault characterises his difference from the Frankfurt School in terms of his problem with these oppositions between life and concept, freedom and power:

> Simplifying things, one could say ... that the conception of the subject adopted by the Frankfurt School was rather traditional, philosophical in nature ... I don't think that the Frankfurt School can accept that what we need to do is not to recover our lost identity, or liberate our imprisoned nature, or discover our fundamental truth ... [but] to move toward something altogether different ... to produce something that doesn't exist yet, without being able to know what it will be. *(2000c:275)*

In place of the need to liberate 'what we really are', Foucault posits a need for 'a total innovation' that would be a *destruction* of what we are as well as a creation of something completely new. Foucault is strongly opposed to any nostalgia or romanticism, either about the past or about an autonomous subject that confers meaning upon the objective world (experiencing that world). The creation of an alternative to present power, objectivity and history can only be achieved within and through power, positivity and history. I would maintain that Foucault and Walter Benjamin occupy a similar position with respect to the critical project of the more prominent representatives of the Frankfurt School.

Concluding 'the critique of modernity'

As a critical account of modernity, Foucault's theory of biopolitics should be seen as a pluralization of Marxist and Weberian critique, not as a substitution or alternative. It calls those perspectives radically into question *insofar as* they are singularising or totalising in their approach, or ground critique in metaphysical assumptions rather than historical events. The theory of biopolitics (contra some of its more vocal advocates) does not replace past epochal accounts of modernity, or ahistorical critiques of rationality, with new improved versions thereof. Biopolitics is not the truth of Western history or of an age. Biopolitics is a programme, a formulation of reasons, experiences, values, embodiment, political process and institutions which work alongside – often in combat with – others, including those that Foucault describes as 'Discipline' and 'Sovereignty' as well as profit accumulation, monopolisation, instrumentalisation and so on. The biopolitical critique of modernity *pluralizes* political economy – placing race, sexuality, health, autonomy and security at the centre of critical concern. It also highlights an 'affective', processual, vitalist, subjectifying, intensive, modern rationality, which is in addition to the modern rationality of classificatory order, mathematical abstraction, normalisation, objectification and extension. We will now turn to the notion of experience in Foucault.

Experience

Experience has been a key, although highly contested and contentious, concept in the critique of modernity in European and American sociology, philosophy, political theory and art since the nineteenth century (Jay, 2006). Romanticism, vitalism, phenomenology, pragmatism and humanist Marxism have all decried an absence of experience in modernity and attempted to recover, liberate or create authentic experience which modern western rationality, intellectualism, industrialisation or commodification have destroyed. In sociology the search for experience is generally associated with an ethos of humanism, an emphasis upon the importance of the subject, and with qualitative methodologies, ethnography and understanding. Experience, in these discourses, is seen as that which confers meaning upon, or grants access to the meaningfulness of, the world. Thinkers from across the political spectrum have treated experience as its own end, advocating a 'politics of experience' or celebrating religious or aesthetic experience. Experience is associated with depth, dimensionality and a complexity of understanding. Experience can be the repository of historical lessons, the sense of connection to the world or the beyond the world. Experience is associated with emotion, embodiment and aesthetics as well as with empirical science.

As Martin Jay notes, Foucault is not often associated with the concept of experience. Foucault is avowedly anti-humanist, opposing the valorisation of the subject and explicitly defining his intellectual project *against* humanist-Marxism and phenomenology (Jay, 2006:360–6, 390–400; Foucault, 2000c). Foucault associated his work with that of Louis Althusser, who was engaged in a protracted debate with English humanist-Marxists over the ideological character of experience (Jay, 2006:190–211). He was cited as an inspiration in various post-structuralist challenges, such of that of feminist historian Joan Scott (Ibid.:249–55), to the treatment of experience as primary, self-evident, or foundational to knowledge. Certainly Foucault would have endorsed Scott's assertion that it 'is not individuals who have experience, but subjects who are constituted through experience', as well as her call to historians to focus their efforts upon the 'analysis of the production of knowledge itself' rather than the reconstruction and communication of lived experience (Scott, 1991:779; 797). Foucault is rightly associated with the rejection of most of the things with which the search for, or celebration of, experience is usually associated. Foucault's work is strongly opposed to the valorisation of 'everyday life' or 'real life' experience, as well as to the kind of epistemology of standpoint that informs identity politics. However, as Jay argues, Foucault, like Walter Benjamin, Georges Bataille and Roland Barthes, can rightly be identified with the 'post-humanist' search for experience without a subject (Jay, 2006:390–400).

Foucault explicitly characterises his genealogical work as history of experience (e.g. 2008b:5–7; 1985a:4). He is interested in the historical formation and creative generation of experience. He does not see experience *as* history, as what happens to an experiencing subject through time. Rather he sees experience itself, the process of *experiencing* – perceiving, seeing, judging, feeling, knowing, being subject – as historical. The organisations of time and space, perception, fields of visibility, perspective, dimensionality and of verifiability are all a part of the constitution of experiencing. All, in Foucault's schema, fall into the category of knowledge, which is historical and discontinuous. Whereas humanistic conceptions of experience understand it as what happens to, or is had by, a subject who brings the capacity to interpret and to create to the experiential encounter, Foucault conceives of those capacities (creation, 'interpretation') as themselves a part of historical experience; not only the object, but also the subject is created in experience. Experience is not the seeing and feeling that is done by a subject, but what is seeable, what is sayable, what is affectable. The history of bodies and of truth is the history of the formation of experience.

Foucault refuses the distinction, which is common to sociological thought, between theory and experience, or the concept and life.

Deleuze argues that Foucault's 'great achievement' was to displace the anthropology of the subject (common to phenomenological, transcendental and humanist investigations of experience) with epistemology (1988:90). This is not to say that Foucault reduces the rich diversity of experience to the intellect, but rather that he sees knowledge as a kind of general organiser, distributing dimensions, fields of visibility, horizons of the sayable and –crucially for the current study – embodiment and flows of affect. Following George Canguilhem, Foucault is interested in the concept as an aspect of living, the 'concept in life'. He writes:

> Forming concepts is a way of living not of killing life; it is a way to live in relative mobility and not a way to immobilize life; it is to show, among those billions of living beings that inform their environment and inform themselves on the basis of it, an innovation that can be judged as one likes, tiny or substantial: a very special type of information. (Foucault, 2000b:475)

Experience, then, is not the opposite or alternative to theories, concepts or intellectualisation. Nor is experience authentic 'real life', nor the radical particularity of 'everyday life' or 'lived experience'. The French term *expérience* has a dual sense, meaning both 'experience' and 'experiment'. Foucault's understanding of experience incorporates something of the alternative sense of the French term, such that experience is an experiment in living, testing the limits of the liveable, creating something new, as well as incorporating abstraction into life, mobilising living.

Benjamin, vitalism and the transformation of modern experience

To develop our grasp of the politics of experience in Foucault's analysis it might pay to pause and consider the relationship between Foucault's thinking and the more established theorist of modern experience – Walter Benjamin. Foucault's interest in the political history of experience shares considerable ground with Benjamin's critical analysis of the transformation (destruction and invention) of experience in modernity. As Howard Caygill has shown, Benjamin's history of technology is the history of the organisation of experience (Caygill, 1998). Benjamin is fascinated by the consequences for experience of transformations in the movements, durability and distribution of objects, materials and physical spaces. The focus of his work on the historical transformation of perception, is the experience of the city and the consequences of technological reproducability. These transformations in perception change the types of values and evaluation that are expressed in and motivate art and politics, tending towards a treatment of life and creativity as values, values that are oriented upon the passing away of time; vitalism.

Benjamin is associated with the critique and history of experience in modernity, with art, with technology and with the transformations in perception – including political and aesthetic sensibilities – that are the outcome of the technological and architectural construction of the modern, capitalist, world. As I have argued elsewhere, Benjamin's interest in these changing perceptions and sensibilities was animated by a concern to develop a historicising account of the vitalist values, and associated search for 'lost experience' (in duration), with which he was contemporary – a vitalism represented in the then 'towering figure' of Henri Bergson (Benjamin, 1999b; Blencowe, 2008). Benjamin's critical engagement with Bergson resonates with contemporary debates surrounding the metaphysical (or not) character of the vitalist philosophy and politics associated with quasi-Bergsonist Deleuze.[2]

Benjamin adopted many of the concepts and values of Bergsonist vitalism, equating communism with a kind of anarchist open state of becoming loose mass, actualising qualitative differentiation in present time. His concept of *Erfahrung* is defined in relation to Bergson's concept of experience in the *durée*/duration.[3] However, Benjamin (in a move that prefigures Foucault's archaeology of biology) sought to historicise the ontologies, values and political aspirations of vitalist philosophy. He argued that the idea of a 'creative' evolutionary life force or *'élan vital'*, that inspired Bergsonist vitalism, as well as the longing for experience in duration, was an artefact of historical transformations in productive technologies (and thus structures of experience) not the expression of metaphysical or timeless truths or values (Benjamin, 1999b). Bergson's *metaphysics* formally posits the *durée* as immanent to the material universe. In his socio-biological account of human experience, however, he *treats* the *durée* as though it were outside of history, as a constant to which history and human experience only change human access (Bergson, 1935). Creativity, the actualisation of the *durée,* appears in this account as the manifestation of the eternal or at least external, a force *flowing into* human life through individual mystic inspiration. In contrast, Benjamin's histories of experience treat the *durée* as immanent in a properly historical sense. For him creativity does not 'flow into' actual human life (as suggested in Bergson's socio-biology) but is the actualisation of virtual difference that is generated and delimited by given *socio-technological conditions*. Modern creativity and vitalism are not the gift of God or the universe but of the capitalist mode of production.

Crucially, Benjamin's fully immanent conception of duration and creative force or life underpins a set of political aspirations that are considerably more realistic and more faithful to the value of 'difference in itself' than is Bergson. Bergson's theorisations of the actualisation of creativity in human experience buy into the navel-gazing notions of romantic individualism. Given the experiential inadequacy of finite individuals it is no surprise that he *effectively* (if not formally) re-posits a transcendent by treating the *durée* as ahistorical. This subordination of difference in itself is engendered in an aspiration for the instantiation of intrinsically authoritarian structuring of experience through the charismatic leadership of heroic mystics, or the enlightenment of philosophers and scientists. In contrast, Benjamin locates the virtuality of creativity in technology and forms of sociality, specifically in the loose mass. Sociality, as action in 'the public', *is* self-sufficient and has no need of a transcendent beyond. Benjamin's aspirations pertain to a proliferation of politics, in something like the Arendtian sense of that term, and thus to a destruction of relationships of domination and servitude (even by and to truly loving heroic leaders).[4] This would mean a radical democratisation and deterritorialization of power, by which the impetus to effect, to create, would be intensified and satisfied in the perpetual, spiritual struggle of and for political and aesthetic practice in the loose mass. Benjamin's redeployment of Bergson's conception of creativity implies the reassertion of the value of difference in itself in the form of the intrinsic creativity of *everyone's* impetus and capacity to *constitute a difference*...in themselves, their environment and other people. It is a call not for love, spiritual renewal and heroic mysticism but for ethics, the politicisation of art and anarcho-communism.

Benjamin, Foucault and the historical production of life

Benjamin's analysis opens up the problematic of experience as historical and plural. His writings invite us to consider material and technological processes through which the dimensionality of experience is produced – productions of qualitative differentiation, constituting *durée*. Crucially, Benjamin contends that the *durée* itself is historically variable. The *durée,* imbedding and investment, need not be about an ancient past or an eternal form, as is assumed by conservative theories of experience. Experience is about a passage through differentiation, through difference, across limits, not through time. Experience as modern creative life, the new *Erfahrung*, is constituted in the present moment, through the proliferation of contingency, connectivity and qualitative transformability. Likewise, as we will see, in Foucault's work the experiential economy of biopolitics pertains not to the conservation of a pre-given natural order, but to the transformative flow and flux of evolutionary transformation, self-transcendence and limit experience, to vitality.

For Benjamin, experience in *durée* might be an incantation of eternity or past time captured in the material configurations of connection that make up tradition or it might be present innovation in open sociality. In Benjamin's view the former has been rendered impossible by the technological conditions of industrial modernity. The introduction of industrial reproducibility, mobility and abstraction into the relationality of objects, detaches experience from the *durées* of past-time and eternal forms. However, it *creates* a new, virtual, dimensionality and depth – *durée* – that can be created in a present moment: a depth, an expansion of present duration, that is constituted through a *politicisation of present relations*, a proliferation of contingency. Industrial modernity and technological reproducibility create a virtual situation in which people open into each other in a loose, qualitative, sociality – influencing each other and being influenced, creating an aesthetic depth through political action rather than tradition. The ontology, politics and ethics of life is, according to Benjamin's analysis, the outcome of these historical transformations in the structure of experience. Vitalist values, celebrating life or creativity, express a kind of (potentially virtuous) necessity of life lived in radical finitude, or radical contingency. In the following chapter we will see that Foucault attributes a similar role to biological knowledge as that which Benjamin does to modern technology. For both, vitalist values express a certain historical, plural, truth.

Whilst arguing that the life of Bergsonist philosophy is a historical production, Benjamin nonetheless advocates a kind of vitalist politics: a politicisation of art, a politicisation of perception, which would actualise the virtual creativity of the second technology and the new *Erfahrung*. Because his conception of creative life is grounded in historical material account of its production, however, Benjamin's politics of life is a politics of political and material *conditions* not a eulogy of a particular conception of real, authentic or free life. Benjamin strives for an anarcho-communism, in which power is perpetually shifting and monopolies are impossible, which would be allied to a radical pluralization and politicisation (making contested) of perspectives. What is envisaged is a production and pluralization of power and materiality, *not* life escaping from power (or matter, or intellectualisation, or the population, or the Empire, or zoēfication of politics...). Again there is a strong resonance between Benjamin's approach to the politics of life and Foucault's. Foucault can be seen to be engaged in an effort to nei- ther negate nor affirm but to *pluralize* politics of life through the work of genealogy and positive critique, whilst his thought is certainly orien- tated towards a work upon the conditions, the political economy, of life rather than its specification, definition or defence.

Benjamin and Foucault both address technologies of reproduction that give rise to contingency and openness – or politicisation – which constitute flows of differentiation, self-transcendence, affect, vitality, embodiment. This productivity, producing life, is not, then, constrained to a force of the somatic or to a vital nature from which either objects or concepts can be excluded. The vital embodiment that is constituted in modernity – through the technological reproducibility of works of art or through bourgeois sexuality (as will be discussed in Chapters 1 and 2) – cannot be identified with either the organic or the somatic. It is both more and less objective, more and less ideal; it pertains to culture and creativity as much as to 'the biological' as that term is usu- ally understood. In Chapter 1 we will see that this term means some- thing very different – something much more historically specific – in Foucault, and that it upsets, rather than being defined by, the culture/ biology, nurture/nature dichotomies.

The historicism about vitalism that Foucault and Benjamin share, and which separates Benjamin (the historical materialist) from Bergson (the metaphysician) does in fact separate Foucault from a great deal of the currently influential literature on biopolitics. In Chapter 3 we will see that Foucault's historicism about life differentiates his conception of biopolitics from that at play in Agamben's thought. In Chapter 4 we will see that the conception of biopolitics as experience – experi- ence that incorporates economies of objects and culture, not simply the somatic – separates Foucault's interpretation of biopolitics from that of Rose. Notably this historicism and pluralism about life (as positivity, as reality and as value) differentiates Foucault from (at least some major incantations of) Deleuze.

Experience, limits and modernity

There are strong parallels between Benjamin's analysis of the transfor- mation in modern technology (organisation of experience) producing a kind of necessary vitalism and Foucault's analysis of the transformations in experience that give rise to biopolitical rationalities and values.

There are significant differences between their approaches however. In Foucault *knowledge* plays the role of 'historical-organiser' that is attributed to technology in Benjamin (Foucault, 2000g:331). Whereas Benjamin is most concerned with experience as culture and life in the city Foucault is most interested in experience as science, reflecting the difference between the French and Germanic approach to the critique of Enlightenment (Foucault, 2000f:440–1).

There is a more fundamental difference between Benjamin and Foucault that pertains to the character of historical transformation. Whilst Benjamin might not be a traditional Marxist his approach to history is far more epochal and dialectical than is Foucault's. Although Benjamin rejects humanist and vitalist nostalgia for 'lost' experience he does participate in the characterisation of modernity as the loss, or destruction, of experience and constructs a dichotomised vision of the possibility of experience before and after modernity. Foucault, always pluralizing, always adding, does not claim that older possibilities of experience have been destroyed with the emergence of new knowledges, nor that only one form of experience is possible in modernity. It would be in line with Foucault's general approach to assume that a variety of types of experience, relating to different regimes of knowledge, coexist in discontinuity.

There is, admittedly, a certain ambiguity surrounding the relationship between experience, as Foucault understands it, and modernity. Experience is tied to life and, in Foucault's view, to the transgression of limits. Both life and transgression are, according to Foucault, historical and associated with modernity. It would be possible to argue, therefore, that Foucault sees experience as exclusively modern – turning all the 'experience-critiques' of modernity on their heads! However, I think that that would be to confuse historical reflection on present experiencing, with an anthropology of the experiencing subject – against which Foucault has consistently railed. Limits and biological life might not be the transhistorical be-all and end-all of experience for Foucault, but they certainly are essential to experience in modernity.

Both Benjamin and Foucault are critical of those nostalgic philosophies of experience in modernity that want to recover some lost, authentic experience of a premodern past. They both associate experience with the allure and epistemic and ethical authority of extending beyond the subject that one is – escaping the condition of finite singularity, becoming differentiated, reaching limits of possibility, participating in process and transfiguration. For both Foucault and Benjamin experience entails not a reassuring embedding of the subject in the world, but passing beyond the limits of and even destroying the subject that one is. For Benjamin experience is either the (now-destroyed) situation in which subjectivity moves outside of the present, into the domain of eternity, or it is the situation of permanent transformation in which people are perpetually differentiated from themselves in the radically open sociality of the 'loose mass' (Benjamin, 2002). For Foucault, experience 'has the function of wrenching the subject from itself, of seeing to it that the subject is no longer itself, or that it is brought to its annihilation or its dissolution' (2000c:241). Experience is 'something that permits a change, a transformation of the relationship we have with ourselves and with the world where, up to then, we had seen ourselves as being without problems ... a transformation of the relationship we have with our knowledge' (Ibid.:244). As such the multiplication of connections, contingency and processuality that is brought about through the technological innovations in material reproduction, or the organisation of embodiment through knowledge of evolution, degeneracy and sexual reproduction, constitute an intensification and multiplication of possible (or 'virtual') experience.

The emergence of modern biology is, according to Foucault, a very fundamental event in the history of experience/truth because modern biology places man 'as a living being' in question. 'By establishing the sciences of life, while at the same [time], forming a certain self-knowledge, the human being altered itself as a living being by taking on the character of a rational subject acquiring the power to act on itself, changing its living conditions and its own life' (2000c:256). We could say that 'life escapes the law of living being' and takes human subjectivity with it in a movement that is radically desubjectifying, pressing against and beyond the limits of the liveable, constituting a plethora, a massification or mass production, of experience.

The history of experience, as an aspect of the history of knowledge or truth, is a part of the history of power for Foucault. The history of experience accounts, in part, for the 'hold' of power. It is a crucial part of genealogical work, not only illustrating that present perspectives are historical, but illuminating, and thus loosening, the grip that they have on present subjectivity. When Benjamin wrote about the possibilities of perception in the context of technological reproducibility he was motivated in large part by the ambition to investigate the appeal of fascism – a growing force at the moment in which he wrote. Fascism, Benjamin argued, provided a substitute for experience through the technological spectacle of mass destruction (1999a:'Epilogue'). Foucault's analysis of biopolitics similarly elucidates the appeal, or affective grip, of eugenicist theory, modern racism, medico-juridical discourse and the theories of degeneracy.

The history of experience is a part of the positive history of power/ knowledge – a crucial, inseparable, component of genealogy. However, experience is also an *objective* for Foucault, and self-justifying end, as it is for humanist critiques of modernity and advocates of a politics of experience. Foucault describes his own work in terms of a pursuit of experience, defining his books as 'experience books' (2000c:243). Of his books he states, 'my problem is to construct myself, and to invite others to share an experience of what we are, not only our past but also our present, an experience of our modernity in such a way that we might come out of it transformed' (Ibid.). He identifies himself with Bataille's and Nietzsche's pursuit of experience where 'experience is trying to reach a point in life that is as close as possible to the 'unliveable', to that which can't be lived through... the maximum of intensity and the maximum of impossibility at the same time' (Ibid.:241; Jay, 2006:390–400). He also identifies himself with a present political struggle for a 'new subjectivity' (Foucault, 2000h:333).

It is when discussing experience as his own objective that Foucault is most explicit about the limit-nature of experience, describing experience as what happens at the limits of what is liveable for a given subject; a 'project of desubjectivation' (2000c:241). It would be easy to insert a false dichotomy here – to assume that limit experience is politically or philosophically 'progressive' or 'good' experience, which could be contrasted with a 'bad' conservative form of experience that is utilised in the grip of the oppressive political discourses of biopolitical modernity. However to make this assumption would in fact be to miss the crux of Foucault's analytics of modern experience. It is, I believe, correct to say that for Foucault *all* experience, or at least all biopolitical experience, is limit-experience, transforming what the subject is, approaching the unliveable, desubjectifying – and this is the case even when a fixed enquiring subject is produced. In particular, the limits of life – death, illness, sex and reproduction – comprise the *experience* of life. In the figure of Bichat, Foucault describes how the experience of life is constituted through the knowledge of death, the 'great white eye' that has seen death 'unties the knot of life' (Foucault, 1973:177). Throughout his career he has taken a particular interest in the ways in which, in Western societies, objectifying encounters with limit-experiences (such as madness, death and crime), knowledge of limit experiences, have constituted people as subjects with a determinate status (Foucault, 2000c:257). As such, my attempt to recover an account, in Foucault, of biopolitical *experience* is addressed to biological knowledge as: the production of life escaping the general laws of being, becoming other than itself; the production of new intensive (densely connected, contingent, politicised) embodiments; and the proliferation of processuality, politicisation and becoming transformative.

Biologism has often been portrayed by sociologists as a form of determinism. The appeal of such determinism, if it is addressed at all, is explained in terms of the appeal of conservatism – the appeal of embedding which is often associated with the idea of experience. In Chapter 5 we will witness Monique Wittig and Rosalind Rosenberg explaining the appeal of Darwinism to 'early' feminists in these terms. The figure of conservative biological determinism is very much alive in contemporary sociology, not as a present form of sociology but as a key character in our past that we use to construct our present. Contemporary biological science is defended from attacks, through arguments that reassure the sociological public that this new biology is about contingency, not determinism, and thus cannot have eugenicist tendencies. More significantly, socially-constructivist ontologies of race, gender and class are assessed, examined for racist, sexist or class-supremacist implications, by way of a comparison with the assumptions of biological determinism – is it reifying? Is it determinist? Does it maintain that essences are fixed?... Foucault's account of biopolitics, as I understand it, suggests that this conservative biological determinist *never existed* – whilst the eugenicist, genocidal, supremacist, regularising socio-biologist certainly did. By situating limit experience within power/knowledge Foucault upsets both the humanistic celebration of the experiencing subject, *and* the positivist (or radicalist) assumption that experience politics is *conservative* politics. Experience in Foucault is about limits, approaching impossibility and intensity. But experience is not only an aspiration of critical philosophy or left-wing politics. Limit-experience is also an aspect of the positivity of biopolitical discourse, a crucial contributor to the 'hold' of biopolitical power. This upsets a number of processes of self-definition and self-assessment that have been important for sociology in the past four decades. It does, as such, pose a genealogical challenge to present sociology, inviting us to reconsider the implications of past biologism for our present. Although this book addresses past history, it is intended as a work upon the present, challenging the existence of a key figure in our present past and thus opening up, perhaps transforming, the (sociological) subjects that we are.

A note on Deleuze

A final 'contextual' issue with respect to the problematic of the book is that of a character that might be conspicuous in his near-absence; Gilles Deleuze. Deleuze's work has been at the centre of debates in sociology and political theory concerning life, vitalism and affective dimensions of power in the past decade. He is associated with a certain 'vitalist' reading of Foucault that is in some respects very close to my characterisation of genealogy in terms of 'positive critique'. Also Deleuze draws together many of the thematic strands that constitute this book, including Foucault, life, vitalism, creativity, temporality, becoming and non-organic embodiment.

Deleuze has certainly been a significant influence in the formation of this study. His *Foucault* has been a trusted and immensely insightful friend as I have developed my own interpretation of Foucault (Deleuze, 1988; also 1995:81–118, 169–76). In particular his comments on 'labour, life, language' and *The Order of Things* was a crucial step on the path to this book (Deleuze, 1988:127, n.10; Foucault, 1970:250–302). The work of a number of feminists who take inspiration from Deleuze have been formative in my thinking on biopolitics, including that of Moira Gatens (1996) and Elizabeth Grosz (2004).

However, as suggested above, there is a crucial antagonism surrounding the place of vitalism in human experience and history, which separates Bergson from Benjamin, and which led me to set Deleuze aside here. The antagonism surrounds the question of whether life, the *'élan vital'* and creativity – which are the subject of vitalist metaphysics, ethics and rationality – are historically constructed formations, or metaphysical, transhistorical forces. Foucault, like Benjamin, comes down very firmly on the historicist side of this dilemma. The *historical* production of life through biological knowledge is a crucial premise and concern for Foucault, and for this book. Deleuze, on the other hand, is genuinely ambiguous on this issue. In *Anti-Oedipus,* written with Felix Guattari, he takes a fairly Foucauldian line, with an intensification of creativity (the proliferation of molecular multiplicity) being described in terms of a historical development and assemblage (2004). Elsewhere however, such as in *Bergsonism* (1991), Deleuze treats vitalism as a matter of metaphysics, not historical analytics. The latter is the less surprising position for Deleuze, who is resolutely a *philosopher* in a sense that is not true of Foucault (the *historian* of thought) (see Osborne, 2003a; 2003b; Blencowe 2008). The historical character of life and biopolitical experience is such a crucial premise for this book that it is appropriate to leave Deleuze, the vitalist metaphysician, in the background. To engage with Deleuze on vitalism would be to engage in vitalist *philosophy more than in the history* of vitalist and biopolitical thought. The topic of this book is the historical production of biopolitical experience, of which vitalism – as experience – is but a part.

Positive critique again

The methodology of the study has itself also been inspired by the idea of positive-critique. The book attempts to conduct a positive-critical reading of social and political theory. This mean expanding upon the force of arguments rather than setting up dichotomising problematisations or demonstrating the fallacy of opposing positions. I attempt to, as Thomas Osborne has put it, 'maximise' the theories and texts under consideration (2008:10). For example, it is not my intention to rubbish or falsify alternative interpretations of Foucault's concept of biopolitics, nor is it to demonstrate the limitations of Foucault's insight. Rather it is to add to, to maximise, the forcefulness of Foucault's texts in the present by illustrating and expanding previously overlooked dimensions of that work. This has influenced the topics and texts that have been included. The intention has been to present alternative positions in the form of productive encounters, to focus upon those aspects of other people's work with which I could *agree* or upon which I could expand, and to develop analyses by way of additions, pluralizations and positivistic descriptions, rather than by negation. In this respect positive-critique means resisting the temptations of the kind of 'point-scoring', so common in theoretical commentary, whereby the authority of the authorial voice is established through its ability to poke holes in other people's reasoning. An ethos of positive critique prioritises the efforts to understand, communicate, translate and interpret over the desire to judge, condemn and denounce.[5] In this respect, as I have discovered in the writing of this book, positive-critique is an inherently difficult, unattainable, problem-generating methodological aspiration. In places it seemed to me impossible to clarify the position that I was myself putting forward without setting up negating critiques of, and dichotomies with, alternative interpretations. For example, I found that in order to communicate both the specificity and importance of my interpretation of Foucault's work I was compelled to provide a negating critique of some alterative positions, so that what is at stake in my own position could be illuminated and the context from which it emerged be explained. Notably I argue (to varying extents) against Agamben's

and Rose's interpretations of Foucault's theory of biopolitics and its contemporary applications. In part this was to remove blocks or plugs that
stood in the way of developing the new account of biopolitics that I
was attempting to draw – contesting, in particular, limiting and reductive accounts of the biopolitical. In part it was simply to explain to the
reader where my own arguments were coming from and to clarify my
position by providing a contrast. Sometimes the duty to communicate,
it turns out, requires the exercise of negating judgement. The final chapter, which most explicitly addresses and defends positive-critique as
methodology of discourse analysis is also that chapter that encounters
the biggest problems with respect to positive-critique as a methodology
for the assessment of theory. The chapter defends the 'positive-critique
of biopolitical discourse' by illuminating the fallacy and danger of the
'negative-critique' that has been prominent in feminist and sociological attacks on biological thought (in particular Wittig, the arch anti-
biology feminist) making for a stark methodological antinomy!

Nonetheless, despite (perhaps because of) the fact that positive-critique is inherently problematic, difficult and unsustainable when it
comes to the explication of theoretical arguments, it remains an expansive, challenging and productive methodological aspiration. Positive-
critique acts as a guiding, unattainable, goal and is realised in the text
as something like an ethos or particular form of integrity, demanding
that we focus our attentions on things we might be able to grasp, that
we *strive* to understand, and that we take up condemnation, judgement
or 'correction' as a regrettable necessity (not as though it were a victory
to be celebrated). It is an invitation to push at and beyond the limits
that we have been given, rather than endlessly precipitating new sets
of limits; an invitation to stretch, pluralize and maximise both differences and the understandings that pertain between them, rather than
to accumulate differences, dichotomies and enmities.

1

Escaping the Laws of Being: The Character of the 'Bio' in Foucault's Genealogies of Biology and Biopolitics

> Perhaps for the first time in Western culture, life is escaping
> from the general laws of being as it is posited and analysed in
> representation ... Life is the root of all existence, and the non-
> living, nature in inert form, is merely spent life; mere being is
> the non-being of life. For life – and this is why it has a radical
> value in nineteenth century thought – is at the same time the
> nucleus of being and non-being: there is being only because
> there is life, and in that fundamental movement that dooms
> them to death, the scattered beings, stable for an instant, are
> formed, halt, hold life immobile – and in a sense kill it – but are
> then in turn destroyed by that inexhaustible force.
>
> (Foucault, 1970:278)

This chapter will give a detailed analysis of what Foucault actually means
by 'biology' in his assertions concerning biopolitics. This demands a
detailed picture of the concept of 'life' that is at play in modern biology. So this chapter sets aside the many issues of *politics* in order to gain
a full and clear picture of what the *bio* in biopolitics actually refers to
(in Foucault's work at least). The key argument of the chapter is that
'biology' and 'bio' in Foucault's work, does not refer to the somatic,
to fleshy living bodies, but to *life* which is in a sense beyond bodies,
existing at the limit of finite bodies, traversing finite lives. Biology pertains to a domain beyond that of individual finite bodies, to processes
that can be described as 'trans-organic', extending beyond the unity

33

and self-persistence of individual organisms. As such 'biopolitics' (in Foucault) is not all and any politics that pertains to peoples' physicality or health, it is specifically the politics of life, of man as a living *species*.

By emphasising this theme of life that is not contained in the limits of living bodies I am highlighting a concern that stretches across Foucault's oeuvre with the limit-nature of life, with experience, and with the role of biological knowledge in the formation of modern political problems. Whereas many commentators divide Foucault's work up into a succession of periods (e.g. Dreyfus & Rabinow, 1983; Merquior, 1985; McNay, 1994), I will be drawing on work from different points in Foucault's career and demonstrating the considerable continuity of his concern with life, limits, experience and politics. These concerns trouble many of the different 'Foucaults'. This is not to suggest that *all* Foucault's work was about a single set of issues, only that some issues, these issues, were an ongoing concern.

Life, as comprehended by modern biology and biopolitics, comes both before and after the bodies of individual organisms, it escapes 'the general laws of being', exceeding the life span of living beings' bodies and thus escaping the ontology of identity. Bringing the trans-organic nature of life into view helps to explain the experiential force of what we might call 'bio-mentality', which is to say: biological knowledge that is an organiser of experiencing, rationality, truth-games, science and embodiment; a horizon of visibility, verifiability and value.

Bio-mentality forms biological, biopolitical rationalities that are antagonistic with the 'rationalism' of classical science and discipline (whilst being, nonetheless, rational). Bio-mentality addresses people as they constitute the life of the species (or population). It does, as such, extend and augment people's embodiment: their capacities to affect and be affected. Like technological reproducibility in Walter Benjamin's analysis of modern experience, knowledge of sexual reproduction, evolution, inheritance and degeneracy transform the durational structure of present experience, introducing contingency, creativity and dimensionality into the present moment. Unlike technological reproducibility, however, biological processes constitute an *immanent* domain of duration, a domain of self-transcendence, of the transcendental – a transorganic embodiment. Life is able to function as a kind of immanent domain of transcendence, producing meaningfulness, value and depth *within* the present (see also Simmel, 1971). Significantly this means that, with biopolitics, the object of power can also become the *objective* of power. In contrast to sovereignty and discipline, biopolitics acts in the name of the very processes that it takes as its object, life.

In Chapter 2 we will see how the trans-organic character of life and embodiment in biopolitics explains the apparently paradoxical formation by which a politics *for* life cannot only allow but place a positive, vitalising, value upon the (political or physical) *death* of living subjects. There is a disjuncture between living bodies and trans-organic biological life. Death, which is the annihilation of living bodies/organisms, can be the very vitality of trans-organic life.

Contesting the culture/biology dichotomy

A common usage of the term 'biological' defines it in contrast to the 'cultural' or the 'social'. Considered thus, 'biology' is tied to one side of a series of dualisms, including physical contra ideal or ideational, and permanent or predetermined contra the historically contingent. In this usage 'the biological' implies materiality and is often associated with the fleshy or genetic body, contrasting with a realm of the mind or of culture. If we are to grasp the force of Foucault's genealogies of biopolitics it is crucial to realise that it is *not* this idea of the biological that is at play in his analyses.

On the one hand 'the biological' in Foucault's account is something much more specific than is that of the biology/culture couplet. Foucault is the archaeologist of modern biology, the science not of living bodies but *of life*, which has a specified point of emergence (in the work of Georges Cuvier) and exemplification (in that of Charles Darwin). When Foucault is talking about biology he is talking about the very specific ways of rationalising, observing and analysing that were developed within modern evolutionary biology and the domains of reality that it addressed (the world as formulated in bio-mentality). As we will see, nineteenth-century biology is addressed to a very different set of phenomena than its predecessor, natural history, although the latter is most definitely also a science of physical-living bodies. The biological, then, is not just *any* matter of living material fact. 'Biology', in Foucault, refers us to a life that is tied to modern formations of knowledge and relationality, which *did not exist* before the nineteenth century (Foucault, 1970:160).

On the other hand, 'the biological' in Foucault's writings is more general than is that of the biology/culture couplet – or rather it contests the limits of those dualisms. 'The biological' refers to forms of relationality, knowing and acting that figure in all manner of domains of practice, many of which could be classified as 'cultural' or 'social'. The biological is not necessarily the somatic. Whilst the biopolitical most certainly *does* include practices that straightforwardly pertain to bodies in their physical-living aspect (to medicine, reproductive control and the like), it also includes a number of practices that have nothing to do with the physical health of organisms. Eugenics at the start of the twentieth century was fundamentally biopolitical. However 'eugenics' described an ethos that was addressed to all manner of phenomena, including principles of literature, architecture, education and design, not simply to reproductive politics. As Mariana Valverde has suggested, significant aspects of Foucault's analysis of biopolitics have greater relevance to contemporary discourses and practices concerning *culturalism* than they do to contemporary medicine or biological science (2007:177).

A genealogical approach to life

As Foucault's archaeologies of the human sciences and medicine demonstrate, life – biological life – is not simply another term for living beings or bodies. Life is something that runs beneath and above, preceding, transcending at the same time as animating, the lives of individual organisms. Life is a distinct and specific reality. It only made its appearance in science, as an object of knowledge and power, at the start of the nineteenth century. Life was not observable previous to this time. Sophisticated centralised technologies of statistical knowledge are required to observe the phenomenon of life. Moreover, observing and affecting life is dependent upon a whole conceptual apparatus concerning reproduction, genetic-inheritance and evolution. Life is something that exists within the inter-generational dynamic time of reproduction, inheritance and evolution. It is not the property of individual organisms.

In order to demonstrate the specificity of biological life, as Foucault understands it, I will also draw upon a history of biology that was written by Nobel Prize winning biologist François Jacob in 1970 – *The Logic of Life: a History of Heredity* (Jacob, 1973). Foucault was a huge fan of this book, describing it as the most remarkable history of biology ever written (Foucault, 1985b:104). Jacob's history of biology concurs with Foucault's archaeology on a number of essential points – in particular concerning the historicity of life as an object of knowledge. His work is a great asset in reading Foucault on biology, adding depth and detail to the history of biological thought which is only schematically articulated in Foucault's own, much wider ranging, works.

Definitions and distinctions

I will also deploy a distinction between 'embodiment' and the 'somatic', using the term 'somatic' (one of the many words meaning body or bodily) to refer to the substantive, physical character and property of bodies. It is suited to describing the contrast between the physically embodied and the cultural or ideal. 'Embodiment' is a much more general category than is 'the somatic'. Embodiment refers to active constellations of capacities. These might extend beyond the limits of a given somatic body or individuatable organism. Embodiment might be, as we will see, 'trans-organic'. Embodiment includes but is not limited to the somatic and organic.

Relatedly, I will argue that it is important to maintain a clear distinction between biopolitics and disciplinary anatomo-politics of bodies. The difference between anatomo- and biopolitics is not simply a question of scale, it is also a question of alternative systems of values and structures of experience and embodiment (forms of vitalism and forms of rationalism). I want to emphasise the trans-organic quality of biopolitical life and embodiment. Biological life is before and after individual living organisms; it is the condition of their existence and it proceeds through their very destruction. Life cannot be reduced to the well-being of living beings. Although the object of biopolitical technologies is often the health of individual bodies, the life that is at *stake* is not that of individuals but that of *populations*. Whilst biopolitics is certainly a politics of the body, of affective force and corporeality, it is not a politics of individual somatic bodies in the sense that discipline is. Biopolitics pertains to a kind of trans-organic embodiment. And it is that very trans-organic quasi-transcendental nature that gives bio-mentality such immense appeal and affective force.

Values and the positivity of life

Further, I want to emphasise the epistemological, aesthetic and affective *positivity* of bio-mentality and to draw out what Foucault's analyses contribute to the genealogy of the *values* associated with life; bio-mentality gives rise to an immense production of experience (of limit experience), an augmentation of capacities, a proliferation of contingency and processuality, a politicisation, an immense positivity. Health, life and security have not always played the central, almost unquestionable role with respect to political and ethical values that they do in contemporary moral and political discourse. These are historical phenomena, contingent upon a range of developments. The knowledge of population life, and the orientation of time upon finitude that it accompanied, are formative conditions of the elevation of mortal life, health and security to a position of ultimate value in political discourse – the position that they still occupy today. That elevation of vitalist, or biological, values is not simply an ideological mask for power. It is a part of biopolitical technologies, including the affective technologies and economy of experience that sustain biopolitical power. The incorporation of individual bodies into biological knowledges and technologies did not just make bodies accessible to power, and expertise accessible to bodies. It also transformed and augmented the present, immanent embodiment of subjects such that these earthly values – health, life, security – could acquire a transcendental character and affective force.

If we do not take on board the depth and specificity of the term 'life' in Foucault's accounts of biopolitics, if we reduce and generalise the politics of life to the politics of living bodies, then the affective force of biopolitics – its force in the production of embodiment, values and desire – escapes from view, disappearing with the trans-organic, self-transcending, quality of life.

Life in Foucault's oeuvre

The question of a specifically modern concept of life, and of its potency as a means for structuring, studying and thinking about human relationships in modern organisation and politics, is present throughout Foucault's oeuvre. The key texts are *The Order of Things* (1970), in which Foucault describes the emergence and defining stakes of modern biology as part of an archaeology of the human sciences; and *The History of Sexuality: 1* (1978) and the Collège de France lecture series of 1975–1979 (2003b; 2007; 2008a), in which he introduces and develops the idea of biopolitics, insisting on the centrality of biological life and epistemology to modern political rationality, values and embodiment. The concern long predates *The Order of Things*, however. The discussion of biology in *The Order of Things* develops key insights from Foucault's earlier archaeology of modern medicine, *The Birth of the Clinic* (1973 [*Naissance de la Clinique*, 1963]). Here Foucault reflects upon the role of medicine as a model throughout the human sciences and draws out the conception of life that is peculiar to this medical model. This medical conception of life is orientated upon a bipolarity of the normal and pathological (1973:41), and harbours death as a continual presence within life (1973:174). He writes: 'societies live [societies are understood to be living beings] because there are sick declining societies and healthy, expanding ones; the race is a living being that one can see degenerating; and civilisations, whose deaths have so often been remarked on are also, therefore, living beings' (1973:41). His interest in the relationship between the specificities of biological life, as modern concept and reality, and modern political technologies (such as modern racism) was, then, already in evidence in *The Birth of the Clinic*.

Much of the literature on biopolitics ignores the earlier figurations of Foucault's concern with the genealogy of life – of *bio* – and its formative role in modern power/knowledge.[1] Foucault's analysis of these themes certainly does shift and develop between these different 'periods'. An interest in the bipolar normalisation of the medical model becomes a more nuanced interest in the dynamic regulation of 'unitary-living-pluralities': of populations. His explanation of the emergence of life (with labour and language) shifts from a discourse concerning epistemic structures to a considerably more historico-materialist account of the development and demands of *technologies* of knowing and power. Nonetheless, key insights concerning the character and capacities of biological life within the relationships and discourse of biopolitics are developed and explained by Foucault in his earlier works on the genealogy of modern biology, modern medicine and the human sciences.

In particular, the crucial distinction that Foucault draws between biopolitics and discipline, in the biopolitics lectures, maps onto the difference that he had already drawn between the classical natural history of the seventeenth and eighteenth centuries and modern evolutionary biology – the science of life – that emerged in the nineteenth century.

For both discipline and natural history, the establishment and exhibition of stable, fixed *order* is a key value. Technologies of knowledge and power include individuation, taxonomy and distribution in two-dimensional axioms (that of the chart or of the factory floor). These techniques contribute to the creation of order as their objective and their means of validation. Both are addressed to a knowable, transparent, observable, orderable world – a world in which divinity (people's and God's) is present in the clarity of extensive order. Both are mechanistic.

In contrast, the science and the politics of life – biology and biopolitics – regard the visible differences, the surface identity of things, as but superficial manifestations, insignificant in themselves. Biology and biopolitics are interested in deeper realities, processes that unfold in hidden depths and across unknowable, enormous, intensities of time – in the duration not of men, but perhaps of Man. The world they address cannot stand still – cannot be arrested in any fixed order, which is by definition a 'mere' artifice. There is an inversion of the location, or source, of truth, value and validation. For biology and biopolitics, normativity and creativity are understood to be manifest *within* the processes of the natural world itself. Life, which biology and biopolitics take as their object, is precisely autogenetic and autonormative – containing creative, evaluative and meaning-making processes within themselves. Power and knowledge are like external managers and observers of these vital, autonormative processes: processes that unfold in the unknown depths of evolutionary time. What really matters is no longer the superficial, easily observable differences between things in the world (and the extensive order that those differences comprise) but a deeper, historic, partially hidden level in which universality is attained not through extension but through identity, singularity, of purpose: the drive to evolution, to reproduction, to life.

Foucault's archaeology of modern biology

We will now turn to *The Order of Things* to develop a detailed picture of the historically specific life – 'bio' – that operates in Foucault's thought. The bio of biopolitics and bio-mentality does not simply refer to the materiality or health of living beings, but to something both more specific and more general. Biological life, according to Foucault, escapes the general laws of being and constitutes a quasi-transcendental domain. It is, as such, a part of the refiguration of history, duration, and experience, transforming the possibilities of affective investment and reason, politicising the present. Crucially for the understanding of Foucault's work on biopolitics, we will see that death and fragmentation are internal to this trans-organic biological life.

In *The Order of Things* Foucault sets out 'an archaeology of the human sciences': of philology, political economy and biology ([1966] 1970). The study details the emergence of these three disciplines at the start of the nineteenth century in Europe, setting them in contrast to their 'equivalent' knowledges in the previous period. It is 'an archaeology', Foucault claims, because it is directed at some deep and forgotten, or unknown, level of scientific knowledge. Addressing three different human sciences at the same time, the archaeology is concerned with those 'network[s] of analogies' and 'rules of formation' of knowledge that are common to all three, relating these also to the philosophies of the time (1970:xi). In contrast to the usual approach of historians of science, Foucault claims to operate at the level of the *'positive unconscious* of knowledge'. The usual approach, he suggests, couples an account of the conscious epistemological level of debates and processes with a search for 'the negative unconscious': the unknown or unacknowledged influences that resist, deflect and disturb science (Ibid.). Foucault is concerned, instead, with 'a level that eludes the consciousness of the scientist and yet is *part of* the scientific discourse': digging up the positive unconscious of science does not mean disputing its validity or seeking to diminish its scientific nature (Ibid., italics added).

The collapse of representation and the insertion of history

Foucault insists upon the radical novelty of biology in the nineteenth century. Its emergence, like that of political economy, is part of a fundamental event in the constitution of knowledge; *the collapse of representation*. Foucault argues that in the eighteenth and seventeenth centuries, the 'classical period' in his schema, biology did not exist. There certainly *was* a science of natural living bodies – natural history, or naturalism – but there was no biology, no *science of life*. Indeed Foucault claims that *life itself* did not exist up to the end of the eighteenth century, 'only living beings' (1970:160).

The collapse of representation was, according to Foucault, a deep event, characterised by a radical transformation not only in relation to knowledge and truth, but in the very structure of experience: the organisation of temporality, dimension and light. The world as apprehended by classical natural history is like a vast, extensive, horizontal plane of all possible difference. It is a world in which *order* provides a sense of epistemic security and meaning (see Foucault, 1970:146). Evidence of such order is found in the total character of the present world, the apparent perfectibility of gradations of difference.

With the collapse of representation this horizontal world is shattered. There is a kind of radical temporalisation – a proliferation of contingency, of the possibility of transformation and of the perception of process – in which the present appears as a passing moment. History displaces order as 'the unavoidable element in our thought' (Foucault, 1970:219). The horizontal order of extension is displaced and taxonomy becomes ordered, instead, in accordance with an 'obscure verticality' (Ibid.:251). An injection of historicity renders present differentiation superficial and scientists begin to refer, for what is important, to great historical forces and hidden depths. Foucault writes:

> Thus, European culture is inventing for itself a depth in which what matters is no longer identities, distinctive characters, permanent tables with all their possible paths and routes, but great hidden forces developed on the basis of their primitive and inaccessible nucleus, origin, causality, and history. From now on things will be represented only from the depths of this density … assembled or divided,

> inescapably grouped by the vigour that is hidden down below, in those depths. (Ibid.:251)

This depth, this new verticality, depends upon so many 'quasi-transcendentals', including life alongside labour and language – 'quasi-transcendentals' that constitute this new depth and plane of history.

Foucault's term 'quasi-transcendental' recalls the transcendental ego, or subject, of Kant's philosophy. The transcendental subject is a kind of universal subjectivity that is somehow before, and external to, experience – to actual consciousness – and yet gives sense to that experience, making it possible *as* experience. Kant's transcendental subject makes experience and understanding possible through the imposition of a fundamental, basic, ordering upon the manifold plurality that is present to the senses – it imposes *a priori* categories such as space, time and causality. It also provides a point of orientation for the exercise of judgement and moral reason – the universalisablity of an action standing as the test of its rightness. Kant's transcendental subject is not transcendent: it is not of another world or even beyond humanity. It is precisely *of* 'Man'. But it is beyond, before, outside of any conscious, experiencing, human – making human experience and understanding possible. The transcendental subject is like intersubjectivity, except it stands outside of history – preceding all experience; a kind of universally given sense, or 'soul' (Schroeder, 2005:18–19).

The term 'quasi-transcendental' is indicative of Foucault's genealogical approach to Kant's philosophy – the *a priori* is, indeed, *a priori* – before experience – but it is also *within* history. The transcendental realm of Kant's universal subjectivity is real, but it is historical – 'quasi'. The 'historico-transcendental' domain is not, for Foucault, Man *per se*, but modern Man – Man with life, labour and language. Life, labour and language constitute a kind of historical *a priori* plane, human, historical, but beyond and before – conditions of – experience and understanding. For Foucault the transcendental is a part of the modern episteme. It is not what makes experience *per se* possible, but what makes *modern* experience, experience as Man, possible – Man that is both the condition and the object of knowledge.

The quasi-transcendentals extend beyond existent being and add a new level, a new depth, to the present moment – the depth in which great historic forces can act and unfold. As such, the quasi-transcendentals take on a hugely significant role with respect to the establishment of truth, epistemic security, organisation of being and relationality in the new disciplines of human science. The quasi-transcendentals are the condition of the new knowledges – knowledges orientated upon history, rather than extensive, universalisable, order. Life, for example, constitutes a unifying orientation of purpose that makes a functionalist, temporalised, knowledge of biological being possible.

Natural history: establishing the reality of order

Foucault defines the specificity of biology against the contrasting, classical, ordering formation of natural history – which is rationalist, like disciplinary power. Natural history, naturalism, or physiology, is the science of natural beings of the classical episteme; of seventeenth- and eighteenth-century Europeans. Foucault paints a picture of the natural historian as the great classifier – the compiler of charts and collector of superficial observations. These natural historians are obsessed with establishing one classificatory schematisation of the whole of nature and they work away at this great chart through the practice of making endless records of visual observations. 'Natural history is nothing more than the nomination of the visible' (Foucault, 1970:132). Rather than considering natural beings in terms of their organic unity, and thus taking interest in the *function* of their organs, natural historians are interested in the visible patterning of the being's organs. So beings are 'paws and hoofs, flowers and fruits, before being respiratory systems or internal liquids' (Ibid.:137).

The object of this great observational and classificatory practice is 'the extension of which all natural beings are constituted', an extension that may be affected by a strictly limited range of variables: 'the form of the elements, the quantity of those elements, the manner in which they are distributed in space in relation to each other, and the relative magnitude of each element' (Foucault, 1970:134). These variables constitute a kind of universal toolkit of observation, such that everyone could and would give the same description of a given entity, whilst any given entity can be considered in a finite set of quantitative relations to all the rest. There is assumed to be a given order in nature, and this order is made evident in the theoretically perfectible gradations of difference that natural historians observe and chart. That order, in turn, holds at bay the radical doubt that Hume introduced into the observation of repetition in experience through pointing to the problem of induction (Ibid.:146). If taxonomy is possible then the order of nature must be more than apparent; taxonomy is evidence that there is a real order to nature.

As is crucial, in Foucault's opinion, this is a *fixed* order; '[t]here is not and cannot be even the suspicion of an evolutionism or a transformism in Classical thought' (Ibid.:150). There is no time *internal to* the order of natural beings. The apparently 'evolutionist' thought of natural historians such as Bonnet propounds a singular movement of nature towards greater perfection. It must, as such, be *preformationist* and thus, according to Foucault, as far removed as is possible from evolutionary thought as we understand it:

> This 'evolutionism' is not a way of conceiving the emergence of beings as a process of one giving rise to another; in reality it is a way of generalizing the principle of continuity and the law that requires that all beings form an uninterrupted expanse...It is not a matter of progressive hierarchization, but of the constant and total force exerted by an already established hierarchy...Such a system, it is clear, is not an evolutionism beginning to overthrow the old dogma of fixism; it is a taxinomia that includes time in addition – a generalized classification. (Ibid.:152)

As has been said, Foucault maintains that life, the object of biological knowledge, did not exist in the classical period. Of course this does not mean that plants and animals were not alive in the eighteenth century; nor does it mean that natural historians failed to note the difference between the living and the not. What it means is that life did not then, as it would for any biologist, mark a radical and obvious cut-off point beyond which an entirely different and specific form of knowledge is required (Foucault, 1970:161). For the natural historians, living beings simply comprised a number of classes amongst the great series of other classes of things in the world. The threshold between what is life and what is not was contingent upon their differing sets of criteria. Maupertuis, for example, defined life by 'the mobility and relations of affinity that draw elements towards one another and keep them together' and thus had to conceive of the simplest particles of matter as alive (Ibid.), whereas Linnaeus only conceived of much more complex organisms as being alive, deploying the criteria of birth, nutrition, ageing, exterior movement, internal propulsion of fluids, diseases, death, and the presence of vessels, glands, epidermis and utricles (Ibid.:160–161). As Jacob puts it:

> Until the end of the eighteenth century there was no clear boundary between beings and things. The living extended without a break into the inanimate. Everything was continuous in the world and, said Buffon, 'One can descend by imperceptible degree, from the best organised animal to the roughest mineral'. There was as yet no fundamental

distinction between the living and the non-living... Organisation still represented only the complexity of visible structures. Throughout the seventeenth century and most of the eighteenth century, that particular quality of organisation called 'life' by the nineteenth century was unrecognised. There were not functions necessary to life; there were simply organs which function. (Jacob, 1973:33–34)

For the natural historians (or 'physiologists'), life was 'a category of classification, relative, like all the other categories, to the criteria one adopts' (Foucault, 1970:161). Life emerges, or 'escapes from the laws of being', with the collapse of representation and the event of biological science. So we have, in the seventeenth and eighteenth centuries, a science of living being that is not a science of life. It is a rationalist and mechanistic science that seeks to uncover the extensive unifying order of the world through the classificatory arrangement of visible structures. It pertains to very different phenomena than does biology and it assumes a very different set of epistemological values.

Biology: life escaping the laws of being

Cuvier and the emergence of biology

Controversially, Foucault identifies the emergence of modern biology with the work of Georges Cuvier, a French natural historian and politician working at the turn of the eighteenth and nineteenth centuries, commonly credited with the establishment of palaeontology (UCMP n.d.). In so doing Foucault is making a number of claims concerning what is essential, at the archaeological level, about the event of biology, claims about transformations in the positive unconsciousness of science that made Darwin's observations and theorisations of evolution possible and appealing.

Within more usual histories of evolutionary biology, Cuvier has been regarded as a conservative thinker, opposing the precursors to Darwin's thought in the shape of the 'early evolutionism' of Lamarck and the like (Outram, 1986:323). Cuvier contested such 'evolutionism' because he maintained that the integral character of organisms was so complete that any transformation in the organism would imperil its survival. However, Foucault insists that these 'early evolutionsims' were not in fact evolutionist at all, in the sense that we have come to understand evolution since Darwin. Their evolutionism was the very opposite of Darwin's because, rather than incorporating dynamism into nature, they added time as an additional component of the *relationally static* taxonomy. Cuvier's treatment of living beings as complete organisms was actually closer to the functionalist evolutionary conception, by which dynamism would be incorporated into relationships.

In identifying the emergence of biology with Cuvier, then, Foucault is confounding the easy and inaccurate dualism of fixism contra evolutionism or progressivism (see Foucault, 1970:150–152). He is contesting the view that what is most essential about evolutionary thought is a belief in progress as a process of perfection, and he is contesting the necessary association between fixism and conservatism. The transformation in the conception of relationality engendered in Cuvier's concepts is, for Foucault, far more essential to the event of biology than is the notion of progress. Or, we could say, the progress proper to Darwin's evolutionary thought is a principle of ordering *within* the world, not (as it is in Lamarck's evolutionism) a metaphysical teleology *of* the world. The stasis that contrasts with evolution, as Darwin understands it, is that of static *relationships*.

Cuvier is famed for the 'principle of the correlation of parts', for proving the fact of extinction and for expounding the theory of catastrophism (including the then novel idea that there has been a time when the earth was dominated by reptiles, who were eliminated in some great catastrophe (UCMP n.d)).

The principle of the correlation of parts is essential in establishing the conditions of possibility of biology according to Foucault. It is the principle that:

> All the organs of one and the same animal form a single system of which all the parts hold together, act, and react upon each other; and there can be no modifications in any of them that will not bring about analogous modifications in them all. (Cuvier, cited in Foucault, 1970:265)

The idea basically consists in viewing each organism as a total and wholly integrated system working toward the singular objective of staying alive. Life is placed at the centre of things. This singularity of purpose forms the core of a functionalist conception of the various organs and systems as integral parts of a unified system – taking us from the natural historians' attention to visible identities and superficial differences to a new focus on depth and function. Foucault claims that from Cuvier onwards:

> It is life in its non-perceptible, purely functional aspect that provides the basis for the exterior possibility of a classification. The

classification of living beings is no longer to be found in the great expanse of order; the possibility of classification now arises from the depths of life, from those elements most hidden from view. (1970:268)

Thus the horizontal plane of differentiation of natural history gives way to a new verticality in the categorisation and interpretation of beings and their components. At the same time the principle entails a radical separation between classes of beings, now considered as complete and completely separate systems for staying alive. Breaking with the assumption of the gradation of differences, Cuvier insists each type of system is complete and thus completely distinct from the rest. For example, being vertebrate is an integral way of maintaining life, wholly distinct from the alternative and also integral way of doing so of invertebrates. As such:

> Whatever arrangement one attributes to animals with vertebrae and those without vertebrae, it will never prove possible to find at the end of one of these great classes, or the head of the other, two animals that resemble each other sufficiently to serve as a link between them. (Cuvier, cited in Foucault, 1970:271)

Demarcations between classes of being are no longer nominal, relatively arbitrary points of classification – they are fundamental disjunctures between distinct, integral, life systems.

As such, a wholly different space of identities and differences is formulated. 'A space without essential continuity. A space that is posited from the very outset in the form of fragmentation' (Foucault, 1970:272). A unification of purpose – of thinking of all organs and organisms as relations to the question of life – blows the natural historians' horizontal gradated order apart, incising fundamental fractures between classes of animals and plants, whilst delving deep beneath the surface of things to expose a depth in which what is most essential and most universal conspires. 'The discontinuity of living forms made it possible to conceive of a great temporal current for which the continuity of structures and characters, despite the superficial analogies, could not provide a basis' (Ibid.:275). The fragmentation of Classical space 'made it possible to reveal a historicity proper to life itself' (Ibid.). In identifying the emergence of the possibility of biology with Cuvier, then, Foucault is stating the importance of fragmentation and disjuncture, including the fragmentations and disjunctures of lives (of death, extinction, destruction) to the emergence of a biological thought that apprehends life itself.

Elsewhere, Foucault emphasises the issue of Cuvier's speciesism (Foucault, 1979). What matters for Cuvier is what pertains to the level of the life of the species. This speciesism was rejected by Darwin, who refutes Cuvier's conception of the type. For Darwin everything happens at the level of the individual organism; the individual's struggle to perpetuate life is the centre of explanation. However, Foucault insists, it is Cuvier's speciesism that engendered the transformation in the epistemic ground – in the positive unconscious – that made Darwin's very individualism possible (Ibid.:128). This alludes to a general theme that Foucault will return to time and again in his studies of biopolitics; the attention modern science pays to the collective life of species, race or population, engenders the possibility (and necessity) of modern individualism. For example, individual sexuality, and sexual self-scrutiny, is of absolute unfathomable importance *because* that sexuality constitutes the *connections* between the individual and the future, past and potential-present life of the entire human species – the exercise of individualism is important because it is affecting collective life (Foucault, 1978:145–156; 2003b:251). In pointing to Cuvier as the starting point of biology, Foucault is insisting on the centrality of speciesism and of attention to *collective life* for modern biology, even to its acute attention to *individuals'* lives.

A final aspect of Cuvier's fame (or infamy) that is telling with respect to Foucault's interest in biopolitics, concerns Cuvier's place within those histories of biology that were written in anti-racist cultural studies and sociology in the 1980s. Here anti-racists such as Sander Gilman and Saul Dubow (working in the tradition of anti-racist sociologists such as Michael Banton) cast Cuvier as a morally despicable, scientific racist, justifying the atrocities that his contemporary Europeans were perpetrating against aboriginal peoples (Kistner, 1999:179–82). There can be no doubt that Cuvier, who credits Caucasians with the entirety of civilisation and refers to Negro and Mongrel people or features as 'disgusting', 'ugly' and 'savage', fits this bill. Although Foucault draws no explicit link between Cuvier and racism, we could say that the fact of Cuvier's racism prefigures one of the central themes of Foucault's later interest in biological-thought-become-political; in racism within biopolitics. Biopolitics, Foucault claims, inscribes racism at the heart of the modern state (2003b:254–258).

The dangers of biopolitics, engendered and enabled by modern racism, are without doubt a central theme of Foucault's work on biopolitics. There is also something telling in the duality of Foucault's and anti-racist cultural studies' attention to Cuvier. It is as though Foucault's archaeology of biology constitutes a prescient critique of the morally outraged, externalist, or 'negative', critique of Cuvier's biology developed within anti-racist sociology and cultural studies. Ulrike Kistner draws out the beginnings of such a critique in his 1999 article. We will expand upon this theme in Chapter 5 when we explore the second-wave-feminist negative critiques of biologism.

In his controversial designation of Cuvier as the emergent point of modern biology, or at least of its possibility, then, Foucault makes a number of claims about the positive unconscious of biology, which will be echoed in his later reflections upon biopolitics. He is confounding simplistic dualisms of fixism contra evolutionism. He is also identifying biology with: a simultaneous fragmentation of the space of being and unification of its purpose; an elevation of themes of death, discontinuity and catastrophe; and with a speciesism that engenders the possibility of individualism. Further, Foucault (perhaps inadvertently) prefigures his later attention to the *racism* of modern-biological-thought-become-political, from which we can draw a more positive-critical alternative to the negative, externalist approach to biology of the morally outraged anti-racist sociology.

Life escaping the laws of being

A distinctive image of life, specific to modern biology, does, then, emerge within Foucault's account of the positive unconscious of the human sciences. He describes the new biological conception of, or attention to, life as 'the escape of life from being' (Foucault, 1970:162). It is different from the objects of the knowledge of natural history. Life as conceived within biology cannot be reduced to that of individual, living, somatic beings. Life is trans-organic, beyond the embodiment of *individual* living organisms. It is beyond bodies because it is quasi-transcendental, outside of, at the same time as within, beings' bodies. It is also beyond bodies because it proceeds through the destruction of such bodies – an animal rather than a vegetable imagining of life.

Foucault conceives of the life of modern biology as a quasi-transcendental. There is a shift from a taxonomic to a synthetic notion of life (Foucault, 1970:269). With this, life becomes something above and below, before and after, as well as *within* living beings (*quasi-transcendental*). It is the *condition* of living beings, not a property of living beings. Foucault writes:

> On the other side of all the things that are, even beyond those
> that can be, supporting them to make them visible, and ceaselessly

destroying them with the violence of death, life becomes a fundamental force, and one that is opposed to being in the same way as
movement to immobility, as time to space, as the secret wish to the
visible expression. (Ibid.:278)

The quasi-transcendental, trans-organic, nature of life is essential to its
epistemic power. Life is posited as the purposiveness of beings: something beyond such beings that they nonetheless constitute. Given this
life, organs and organisms can be understood from the point of view of
function – their functionality for perpetuating life. Such functionalist
explanations constitute the organising principle of categorisation and
explanation within biological thought. Life is the principle or ground
of sense and purpose within the world of living beings as apprehended
by biological knowledge: life is quasi-transcendental (see also Jacob,
1973:88–92).

Death and the destruction of beings' finite lives is inscribed within
the perpetuation of this potentially infinite quasi-transcendental
life. At the level of imaginative values this life would be henceforth
'expressed in the form of animality' (Foucault, 1970:277). Whereas the
botanical garden had served as the archetype of the taxonomic life
known through natural history, biological life is made perceptible in
the destructive drama of the animal kingdom. Resonating with his later
insistence on the complicity between biological values and the administration of death, Foucault writes:

Transferring its most secret essence from the vegetable to the animal
kingdom, life has left the tabulated space of order and become wild
once more. The same movement that dooms it to death reveals it as
murderous. It kills because it lives. Nature can no longer be good.
(Ibid.:278)

Life is not the well-being of bodies. Life is a great, potentially infinite,
current, above and below, within and beyond bodies and their lives.

In relation to life, beings are no more than transitory figures, and the
being that they maintain, during the brief period of their existence, is
no more than their presumption, their will to survive ... individuality,
with its forms, limits and needs, is no more than a precarious
moment, doomed to destruction, forming first and last a simple
obstacle that must be removed from the path of that annihilation.
(Ibid.:278–279)

Biological science is not addressed to the same world as is natural history. The world of biological science, the world as apprehended and organised by biological knowledge, the bio-mental world, is a world of present depth, obscure but knowable forces, fundamental fractures and incessant, intensive processes. Bio-mentality pertains to a reconstitution of temporality and differentiation – of *durée* – of experience and of what matters. As Jacob states, a certain vitalism became as essential to nineteenth-century science as mechanism had been to the eighteenth (1973:92). Functionalist explanations, and later evolutionary theory, posit life and its perpetuation as the defining objective of organic beings. That life stands, if not above then at least on the outside of – before and after – such organisms' existence. The bodies of given organisms halt life, hold it immobile for an instant, 'in a sense kill it – but are then in turn destroyed by that inexhaustible force' (Foucault, 1970:278). What matters, from the point of view of bio-mental rationalities, is not the organism in itself, but the organism in its relation to trans-organic life, life itself, which is perpetuated in sex and in death. Foucault's depiction of biological life, in which death is internalised, prefigures his later analysis of specifically modern racism, which configures 'biological-type relationships': relationships that can vitalise death. The biological-type relationship means that the 'fact that the other dies does not mean simply that I live in the sense that his death guarantees my safety; the death of the other, the death of the bad race, of the inferior race (or the degenerate, or the abnormal) is something that will make life in general healthier: healthier and purer' (Foucault, 2003b:255). Trans-organic life persists through sexual reproduction, through the generation of new life, but also through the death and the destruction of present organisms. We will elaborate upon the biological-type relationship and its thanatopolitical implications in the following chapter.

The persistence of physiology/natural history

Whilst Jacob does not refer to Foucault, his analyses of the history of biology largely support Foucault's claims in *The Order of Things*. In particular, Jacob concurs with Foucault in insisting upon the novelty of life as a concept in the nineteenth century (as we have seen above); upon the primacy of intergenerational concepts – heredity, reproduction, evolution – to the logic of life; in suggesting that Cuvier, rather than the usual contenders such as Lamarck, might be rightly considered the predecessor to Darwin (1973:12–13); and in emphasising the close relationship between transcendentalism and the emergence of life as functional unifier (Ibid.:278).

Whilst Jacob's history of biology is largely in agreement with Foucault's, there exists a significant difference between them. Jacob makes it plain that the tendencies, values and practices that defined eighteenth-century natural history, or (as Jacob terms it) 'physiology', did not disappear with the emergence of modern biology and its concepts of life, evolution and reproduction. For Jacob these two movements of thought represent alternative tendencies – tendencies that have, in fact, been ever-present in the science of living beings (although one was dominant in the eighteenth century whilst the other has been dominant since).

Contrary to what is often imagined, biology is not a unified science... At the extremes *[of the intellectual diversity]* are two great tendencies, two attitudes in fundamental opposition. The first may be called integrationist or evolutionist. Not only does it claim that the organism cannot be separated into its components, but also that it is often useful to consider it as an element of a system of higher order – group, species, population or ecological family. Evolutionary biology is concerned with communities, behaviour, the relations which organisms set up with one another or with their environment. (Jacob, 1973:6)

Foucault's characterisation of biology clearly resonates with Jacob's characterisation of specifically *evolutionary, integrationist* biology. Jacob continues:

In fossils *[evolutionary biology]* looks for traces of present-day living forms. Impressed by the incredible diversity of beings, it analyses the structure of the living world, seeks the cause of existing characteristics and describes the mechanism of adaptations. It aims to define the forces that have made the fauna and flora of today. For the integrationist, the organ and the function are interesting only as part of a whole that comprises not simply the organism, but the species, with its impedimenta of sexuality, prey, enemies, communication and rites... For him, biology cannot be reduced to physics and chemistry, not because he desires to invoke mystically a vital force, but because integration confers on systems at all levels properties which their elements do not posses. (Jacob 1973:6–7)

The second tendency that Jacob describes is more resonant with Foucault's characterisation of the style of study of the earlier natural history.

The opposing attitude may be called tomist [sic] or reductionist. For the reductionists, the organism is indeed a whole, but a whole which must be explained by the properties of its parts alone. He is interested in organs, tissues, cells and molecules. Reductionism seeks to explain functions by means of structures alone. Conscious of the unity of composition and functioning found behind the diversity of living beings, it sees the organism's performances as the expression of its chemical reactions. The reductionists believe that the components of a living being must be isolated and studied under controlled conditions...For him, there is no property of the organism which cannot ultimately be described in terms of molecules and their interactions. Certainly there is no question of denying the phenomena of integration and emergence. Without doubt the whole may have properties which its components lack, but these properties always result from the structure of the components themselves and their arrangements. (Ibid.)

These reductionists certainly do not sound like biologists as Foucault has defined that category.

Drawing from Jacob, we could say that the two types of study of living beings that Foucault characterises participate in opposing tendencies of the science of nature. Whereas Foucault implies that the former, natural history, was supplanted by the latter, (evolutionary) biology, Jacob, more convincingly (and in actually *more* 'Foucauldian' or genealogical fashion), maintains that the former tendency persisted, in altered forms, alongside the new science of life. In the eighteenth century, the mechanistic, ordering, taxonomic tendency was dominant. In the nineteenth, the more 'vitalist', synchronic, integrationist tendency of evolutionary biology was dominant – but it added to and transformed, it did not eclipse, the former.

At the end of the nineteenth century, and for at least the first half of the twentieth, Jacob states, these two tendencies coexisted in relatively hostile, separate and uncommunicative camps; biochemistry and genetics (Jacob, 1973:179–80). On the one hand we have *biochemistry,* which – like eighteenth-century physiology – is mechanistic and taxonomic. (*Unlike* eighteenth-century physiology, however, it is experimental and molecular – as though the emergence of evolutionary biology had forced the search for order onto another plane, digging down beneath the observable surface of things.) On the other we have *genetics,* the science of inheritance, populations and hidden structures. This is the twentieth-century formulation of the science specifically of *life,* the science that we (following Foucault) are defining as 'biology'.

Taking Jacob's distinction into account, we can say that when Foucault uses the terms 'biology' and 'life', he is referring specifically to evolutionary, genetic, biology and the sorts of collective and serial phenomena that it takes as its objects. The taxonomic, physiological, approach to the study of living beings has continued alongside evolutionary biology in numerous guises, including the biochemistry of the first half of the twentieth century.

Interestingly, from the point of view of contemporary developments in the constitution of the 'bio', Jacob explains that biochemistry and genetics did in fact come together and join forces by the latter half of the twentieth century with the emergence of *molecular biology*. With this, 'the two major currents of biology, natural history and physiology, that went their separate ways for so long, almost unaware of each other, have finally joined forces. The old quarrel between integrationists and reductionists has [now] been resolved in the distinction established long ago by physics between the microscopic and the macroscopic' (Jacob, 1973:299).[2]

Foucault's analysis of the crafting, taxonomic, ordering, physiological approach of power and knowledge, typified in natural history and biochemistry, is carried over into his studies of discipline. His analysis of the science of life, genetics and biology, is carried over into his concern with biopolitics. Whilst Foucault's account of life in modern science is quite different from his later account of life in modern government, the former certainly does inform, and give analytical depth to, the latter. The themes touched upon here – of life being beyond the lives of living beings, of life proceeding through fragmentation and destruction, of life as synchronic not taxonomic, and of life as a depth in which both truth and purposiveness are found – these themes are all present in Foucault's genealogies of biopolitics. The genealogies of biopolitics do address a different set of topics than do the archaeology of human science and it would do a violence to either project to collapse them into each other. However, the genealogies of biopolitics are an explicit *development* of the archaeology of the human sciences and they should be read together, at least insofar as they are both studies in the genealogy of life-as-value, questioning what gives 'our' most ultimate, unquestionable value – the value of life – its value.

Foucault's genealogy of biopolitics: managing and maximising life

Having developed a detailed and precise picture of the history and philosophy of biological life that informs and animates Foucault's analysis of biopolitics, the chapter will now move on to consider the biopolitics literature directly. This section will establish that the life that is in play in biopolitics is the same as that in modern biology. As such, biopolitics is not the politics of living bodies but of specifically trans-organic life. Crucially, I will establish a clear distinction between the domains of discipline and of biopolitics – a distinction that is fundamental to the very different formation and content of values that the two technologies of power engender.

Foucault introduces the theme of biopolitics in the conclusion of *The History of Sexuality: 1* (1978) and the final lecture of the *Society Must Be Defended* series at the Collège de France of 1975–1976 (2003b). These are, ostensibly, genealogies of the modern obsession with sex and sexuality, and of the discourse concerning war as analyser of history and society (so the themes are sex and death; the perpetuation of trans-organic life). They both conclude with an extended and rhetorical discussion of the biopolitical, and biopolitically racist, character of modern power. Man's modernity has been reached, Foucault claims, when life – biological life – makes its entry into the field of history; 'into the order of knowledge and power, into the sphere of political techniques' (1978:142). The concepts and archaeology of biopolitics are further developed in the next two lecture series: *Security, Territory, Population* (Foucault, 2007), wherein Foucault draws out links between biopolitics and the 'care for life' of the Christian pastor; and *The Birth of Biopolitics* (Foucault, 2008a), in which Foucault develops an analysis of liberalism as essentially biopolitical. In addition, he discusses them in a number of lecturers and interviews of the period (e.g., Foucault, 2000h; 2000i; 2000j).

In the course summary of the *Birth of Biopolitics*, Foucault defines biopolitics as:

> the endeavour, begun in the eighteenth century, to rationalize the problem presented to governmental practice by the phenomena characteristic of a group of living human beings constituted as a population: health, sanitation, birth rate, longevity, race... We are aware of the expanding place these problems have occupied since the nineteenth century, and of the political and economic issues they have constituted up to the present day. (2008a:317)

The development of biopolitics is bound up with that of 'governmentality', the securitising conduct of conduct, and constitutes a second stage in the development by which new forms of power, of positive productive powers to *make live,* supplemented and supplanted the technologies of sovereignty, wherein power is essentially so many tools of deduction, negation and refusal – powers to *take* life. These events are most clearly set out in the final lecture of the *Society Must Be Defended* series (Foucault, 2003b), which will be the key source for this section.

There is a tendency in the literature on biopolitics to treat discipline and biopolitics as two versions of the same thing: as two modalities of politics of the body. First this equivalence is established through an epochal approach to discipline and biopolitics, whereby a 'disciplinary society' is said to have been replaced by a 'biopolitical- or 'control-society' (Deleuze, 1995:177–82; Hardt and Negri, 2000). This implies a functional equivalence between discipline and biopolitics. Also the term 'society' here implies that both forms of power have a kind of total reach, taking hold of embodiment (at least) in its entirety. Both implications run counter to Foucault's arguments, as we will see below. Second, the equivalence is established, more plausibly, by collapsing the two technologies into each other, treating them as two aspects of the same thing, such that we have a singular anatomo- and biopolitics of the somatic body. This general approach is shared by Agamben (1998) and by Rose (2001; 2007), although they differ very substantially in their accounts of what that politics of the body is. Building upon the above interpretation of life in biology, and drawing on Foucault's writings on biopolitics, I want to insist that discipline and biopolitics are more distinct than Agamben and Rose allow. Biopolitics and discipline are addressed to different modalities of embodiment. They engender *conflicting* values, though do not necessarily come into conflict (as they operate at different levels). They should neither be treated as totalising formations of power, definitive of distinct epochs, nor collapsed into a singular technology of somatic-power. Rather they should be understood as two distinct technologies of knowledge, formations of power and experience, which have been coextensive through the past 200 years. Discipline is a politics of men, women and children as bodies: as individuated somatic organisms. Biopolitics is a politics of and for life, which is trans-organic, transcending at the same time as constituting and being constituted through individual organisms. Biopolitics pertains to *populations* – specifically political trans-organic embodiment – not somatic bodies but autonormative, autogenetic – vital – processes of life.

Biopolitics contra discipline

The *Society Must Be Defended* lecture series explores the genealogy of modern racism, and of the conception of war as analyser of history, through an account of the aristocratic histories of race war (Foucault, 2003b). In the penultimate lecture, Foucault looks forward from the moment of aristocracy to republican modernity and argues that a new grid of historical intelligibility came into play at the time of the French Revolution – a shift away from concern with ancient pasts and original rights towards the present (and capacities within the present) as the fullest, culminating, moment of history. In the final lecture, Foucault draws a schematic account of a new form of racism – state racism – that would be crucial to modern political technologies. To understand the character, significance and, as it were, technological utility of this modern racism, Foucault suggests, we have to understand the broader context of *biopolitics* out of which it emerged. It is his description of that context that will be summarised here.

Foucault opens the discussion with the issue of positive power. From the eighteenth century onwards, new modalities of power were developed that were positive, productive, in a sense that contrasts with the essentially deductive powers of sovereignty. Sovereign power was deductive – the power to take or deny – and it was ultimately manifest in the power of the Sovereign to take the life of his or her subjects. In the nineteenth century, Foucault argues, political right underwent a fundamental transformation. There emerged a new right, which supplements, permeates and penetrates the old right of the sovereign – the right of power to 'make' live and 'let' die. This right is essentially the right of the new Nation States and engenders positive powers that can invest bodies, making them more forceful, more active, at the same time as normalising them (Foucault, 2003b:240–242).

This general theme, the productivity of modern power, is familiar to readers of Foucault from his earlier *Discipline and Punish,* the famous genealogy of the prison. In *Discipline and Punish,* Foucault describes disciplinary power as a form of *investment* in bodies: infiltrating, normalising (making docile); but at the same time energising, economising (making useful) (Foucault, 1977: esp. 135–169; Osborne, 2008:107–113). Famously, Foucault identifies Jeremy Bentham's panopticon prison design as a key diagram for, and exemplification of, disciplinary power. Technologies of disciplinary power were first developed in the seventeenth and eighteenth century – especially in the context of emergent disciplinary institutions; schools, prisons, hospitals and barracks. In *Society Must Be Defended* (as in *The History of Sexuality 1*) Foucault characterises disciplinary power as the 'anatomo-politics of the human body' and describes it as a first stage in the development of positive power. The development of discipline was contemporary with natural history and shares in many of its tendencies. The emergence of biopolitics came at least a century later, at the end of the eighteenth century, and constitutes a second, decisive, stage in the development of positive power: a stage that enables a new right – a new authority – to be established, a right grounded upon the principle of *making live* (Foucault, 2003b:241–242).

The second form of positive power, biopolitics, emerged at the end of the eighteenth century. This new technology dovetails into disciplinary power, but it is not itself disciplinary. It exists at a different level or scale, bears upon a different area, and makes use of very different instruments. 'Unlike discipline, which is addressed to bodies, the new non-disciplinary power is applied not to man-as-body but to the living man, to man-as-living-being; ... man-as-species' (Foucault, 2003b:242). Discipline is individualising – working upon groups of people only to the extent that their multiplicity can be broken down into individual bodies – 'bodies that can be kept under surveillance, trained, used, and, if need be, punished' (Ibid.). The new power in contrast is addressed to groups of people insofar as they form a 'global mass', a collective embodiment that is affected by overall processes:

> So after a first seizure of power over the body in an individualising mode, we have a second seizure of power that is not individualising but, if you like, massifying, that is directed not at man-as-body but at man-as-species. After the anatamo-politics of the human body established in the course of the eighteenth century, we have, at the end of that century, the emergence of ... what I would call a 'biopolitics' of the human race. (Foucault, 2003b:243)

Foucault goes on to enumerate the interests and the domains of application of this biopolitics. Biopolitics is closely tied to the development of statistical knowledge, addressing phenomena that are apparent at the statistical level. The things that it takes as its objects (of both knowledge and intervention) are serial collective phenomena, such as the birth rate, the mortality rate, various biological disabilities and the effects of the environment. According to Foucault, we see, at the end of the eighteenth century, the beginnings of a natalist policy with the development of plans to intervene in all phenomena relating to the birth rate (a general plan that was widely effected in the nineteenth century). We see a new interest in the impact of death and illness on populations – their impact as ever-present, ever-influential, factors that might be controlled. We see a new interest in various forms of incapacity, such as old age or infirmity, and the development of new strategies to deal with these – not only the augmentation of charitable provision (which had been in existence for a long time) but the introduction of novel, more subtle and rational mechanisms such as insurance, individual and collective savings, and safety measures (Foucault, 2003b:243–244).

Biopolitics is not interested in phenomena insofar as they impact upon individuals, as bodies or otherwise. It is interested in a new – newly knowable – type of embodiment, that of the *population*. Which is to say:

> a multiple body, a body with so many heads that, while they might not be infinite in number, cannot necessarily be counted. Biopolitics deals with the population, with the population as a political problem, as a problem that is at once scientific and political, as a biological problem and as power's problem. (Foucault, 2003b:245)

This body is not that of society – at least not as 'society' was understood by the theorists of right (which is to say as the collective interests and actions of individuals). As Foucault puts it elsewhere '[t]he social "body" ceased to be a simple juridico-political metaphor (like the one in *Leviathan*) and became, instead, a biological reality and a field for medical intervention' (2000j:184).

Biopolitics is concerned with issues that are only relevant at a mass level – rates of birth and death, illness, insurance – issues that are irregular, unpredictable, ungovernable at the level of the individual. It is addressed to serial phenomena: phenomena that only exist across significant periods of time.

> The mechanisms introduced by biopolitics include forecasts, statistical estimates, and overall measures. And their purpose is not to modify any given phenomena as such, or to modify a given individual insofar as he is an individual, but, essentially, to intervene at the level at which ... general phenomena are determined, to intervene at the level of their generality. (Foucault, 2003b:246)

Foucault is keen to stress that biopolitical technologies are *not disciplinary*. Whilst biopolitical and disciplinary mechanisms do share in the objective of maximising and extracting forces, and are both technologies of the body, they are addressed to very different forces and to very different formations of embodiment. Biopolitical mechanisms do not work at the level of the body itself. It is a matter of using *overall* mechanisms and acting to achieve overall regularity:

> One technique is disciplinary; it centres on the body, produces individualizing effects, and manipulates the body as a source of force that has to be rendered both useful and docile. And we also have a second technology which is centred *not upon the body but on life*: a technology which brings together the mass effects characteristic of a population, which tries to control the series of random events that can occur in a living mass, a technology which tries to predict the probability of those events (by modifying it, if necessary), or at least to compensate for their effects. (Foucault, 2003b:249 *italics added*)

The distinction between addressing the bodies of living beings and addressing biological life itself – life that escapes from being – that Foucault developed in *The Order of Things,* is clearly in play in this distinction. Discipline is a politics of individualised living bodies, biopolitics is a politics of *life*.

So what *does* biopolitics do if it does not discipline? The key terms that Foucault repeats are 'regularising' and 'securing'. 'It is, in a word, a matter of taking control of life and the biological processes of man-as-species and of ensuring that they are not disciplined but regularised' (Foucault, 2003b:246–247). Biopolitics installs security mechanisms around the random element inherent in a population of living beings 'so as to optimize a state of life' (Ibid.:246). It is a technology that 'aims to establish a sort of homeostasis, not by training individuals, but by achieving an overall equilibrium that protects the security of the whole from internal dangers' (Ibid.:249). Biopolitics will regulate variations, eliminate the random element, support *existing* processes and it will act upon an interest not in individual organisms but in collective life.

Whereas discipline is creating an artificial order that gains its validity, its, as it were, 'reality principle', from the strength and clarity of that order, biopolitics is *managing an already existing process,* a process that has its own validity, its own reality, its own normativity. Biopolitics identifies and fosters processes. It enables processes to become what (it has been decided that) they already, really, were. We could say that discipline creatively *imposes* a (set of) diagrammatic norm(s) upon individual bodies whilst biopolitics managerially regulates, fosters and secures the *auto*normative, autogenetic, processes of biological life.

Foucault suggests that we regard discipline and biopolitics as two stages in the progressive response, on the part of power, to the inadequacy of sovereignty as a technology of power in the face of a demographic explosion and industrialisation from the seventeenth century onward. First, power was adjusted to take care of the details that were escaping the mechanisms of sovereignty. Disciplinary mechanisms were developed in the seventeenth and eighteenth century, particularly within the framework of institutions such as schools, hospitals and barracks. Second, a more difficult, complex, adjustment was made; an adjustment of power to the phenomena of populations, 'to the biological or biosociological processes characteristic of human masses', actualised primarily in the context of the State (Foucault, 2003b:250). Biopolitics does not supplant discipline, any more than discipline supplanted sovereignty and the law. These different technologies of power do different things. They have conflicting values, but often this conflict remains unactualised because they are acting upon and in different realities, different domains. Sovereignty, discipline and biopolitics coexist in a 'triangle' of technologies of power in the context of modernity (see Foucault, 2007:108).

Foucault explicitly acknowledges that this interpretation of the delimited nature of discipline, and of the role of biopolitics in the nineteenth century, contradicts the position that he had taken in *Discipline and Punish* ([1975] 1977), which remains his most well-known book amongst social scientists. In *Discipline and Punish* Foucault does characterise the nineteenth century as 'disciplinary' and implies that disciplinary power had taken hold of the totality of bodies. In *Society Must Be Defended,* however, Foucault explicitly rejects the idea that the normalising society of the nineteenth century is one in which disciplinary institutions have swarmed and taken over everything – this, he states, was 'a first and inadequate interpretation' (2003b:253). Discipline and biopolitics can both be understood as normalising technologies, but they activate normalisation in entirely different ways (Ibid.). As has been said, discipline *imposes* a (set of) diagrammatic norm(s) upon individual bodies whilst biopolitics regulates, fosters and secures the *auto*normative – selforganising – processes of biological life. A normalising society is one in which (at least) two very different types of normalising power operate: imposing norms *and* fostering autogenetic, autonormative processes of population; disciplining living bodies *and* fostering biological life.

A crucial point to draw out is that biological life is not only the object of biopolitical power, it is also that power's *objective*. Biopower *is* 'the power to guarantee life' (Foucault, 2003b:254). Biopolitics 'takes life as both its object and its objective ... its basic function is to improve life, to prolong its duration, to improve its chances, to avoid accidents, and to compensate for failings' (Ibid.:254). Biopolitics is not some ideological disguise for a sovereign power that is still bent upon the augmentation of the power or wealth of the sovereign. For Foucault, the objective, the internal value, the motivation, of biopolitical power *really is* the maximisation of population life – a maximisation that is not disciplinary and not simply exploitative, but is also fostering, caring, securing. Life can be the objective of biopolitical power because it is, in itself, a field of utility, value and veridification. Life becomes its own end in the context of biopolitics.

The biopolitical values of life as an end in itself – vitalist values – become so powerful in nineteenth- and twentieth-century Europe that they are central to the political aspirations and efforts not only of State power, but also of oppositional movements and challenges to the State (Foucault, 1978:144–145).

Disciplinary power exploits, orders and invests bodies. It increases the forces inherent in bodies and it implants new ones. Those bodies that form the object of disciplinary power are not its objective. Discipline refers for its order and validation to something outside of its object. It crafts, produces, designs: the sense, value and meaning are drawn into, inscribed upon, the object from elsewhere.

Discipline and biopolitics are not two versions of a politics that takes the body as its object. They are two very much distinct formations of power that take different objects (the individual as body or the vital processes of the collective body), and they are animated by very different formations of value (aiming at the artificial construction of meaningful order, or aiming at the maximisation and regulation of life processes, which are in themselves meaningful, ordering and creative).

Conclusion

It is clear in Foucault's account that biopolitics is not simply the politics of bodies or health – any more than biology is simply the study of living beings. Just as he explains the emergence of biology (and the autonomisation of its object, life) in terms of a radical event in the organisation, knowledge and imagining of relationality (and thus temporality), so too does he posit a transformation in relationality as something like the kernel of biopolitics. Biopolitics is what emerges when man starts to understand that 'he is a living *species* in a living world' (Foucault, 1978:142, italics added). Biopolitics does not simply come into view in investigations into the history of medicine; Foucault develops the idea in his genealogies of sexuality (1978), of modern concepts and technologies of class, race, society (2003b), and of civil society (2007). It seems, then, that biopolitics is engendered in the practice of paying attention specifically to trans-organic life and inheritance, rather than to the specifically somatic.

Biopolitics is a form of governance that is addressed to a domain of reality which, if not exactly invented, was brought into being as an object of knowledge and technique, through numerous scientific and strategic innovations that clustered around a concept that had developed a wholly new sense in the eighteenth century – *the population* (Foucault, 2007:245). Biopolitics is the politics of the life of the population.

> Previous to the eighteenth century 'population' usually meant the process that is the simple inverse of 'de-population'. Where it was a noun it would name the collection of individuals, the subjects, living in a given territory. As statistics developed vital serial phenomena became observable at a societal or national level and the population was 'discovered' or 'invented' as a vital existent in itself, with its peculiar collective and serial characteristics and properties. (2007:79)

Population is not exactly an idea or invention. Nor is it exactly a 'discovered object'. Foucault claims, instead, that it is a domain that is carved out through an interplay of power and its object. He writes:

> A constant interplay between techniques of power and their object gradually carves out in reality, as a field of reality, population and its specific phenomena. A whole series of objects were made visible for possible forms of knowledge on the basis of the constitution of the population as the correlate of techniques of power. In turn, because these forms of knowledge constantly carve out new objects, the population could be formed, continue, and remain as the privileged correlate of modern mechanisms of power. (Ibid.)

The population is defined not by its specifically physical aspect but, rather, by its connectivity. In this new population each individual is seen to have an autonomous and spontaneous bond with all the others. The population is comprised of discreet but connected and interdependent bodies – but it is more than, different to, those individual bodies. The population constitutes not a collection of subjects but a set of natural phenomena with its own immanent laws of transformation and movement (Ibid.:351–352).

Whilst it is not a term that Foucault uses, the idea of 'trans-organic embodiment' is appropriate to grasp the domain of population life. Population is, as we have seen, trans-organic – persisting through the very demise of individual somatic organisms. It is, nonetheless, appropriate to conceptualise the population and its life as a form of embodiment, a 'new body, a multiple body' as Foucault puts it. Whilst the population is not exactly an organism, it is an organisation of affective force, bodily influence and bodily flows – flows of fluids, genetics and inheritance.

By the late 1970s, Foucault has come to regard the emergence of this trans-organic embodiment, the population, as the key element in the formation of the positive unconscious of modern science and human science (in addition to modern politics). This follows a movement in Foucault's approach to history such that politics, rather than epistemology, becomes paramount in his analyses.

Foucault returns to the labour, life, language triplet of *The Order of Things* in the *Security, Territory, Population* lectures. Here the three are not characterised as quasi-transcendentals but are, instead, referred to as 'a domain'. Displacing the theory of epistemic failure in the collapse of representation, Foucault incorporates life, labour and language into the historical emergence of population. The population is the 'operator' of the transition from natural history to biology, from the analysis of wealth to political economy, and from general grammar to historical philology – which is to say, of the emergence of life, labour and language (2007:78). To some extent, then, the archaeology of the population can be seen to supplant – not simply to supplement – the archaeological work of *The Order of Things*. The shift from epistemology to politics – from the history of the human sciences to that of governmental rationality – is not simply a shift in topic; it also appears as a profound shift in Foucault's approach to the work of genealogy – to the historical account of the production of values and knowledge. We might say that Foucault has become more sociological, or more materialist, moving from problematics of philosophy to those of relationships, organisation and political technology.

Foucault does not refer to life as a 'quasi-transcendental' when on the topic of biopolitics. However, the self-transcending, immanent character of a quasi-transcendental trans-organic life seems to be manifest in something like a material or pragmatic form in the life of the population. In the population, Foucault has identified the more empirical, more material, ground of the arbitration that he once approached in the epistemology of quasi-transcendentals. The life of the population is still the life beyond somatic bodies described in *The Order of Things*. It is a collective life, active and affected at the point of inheritance between generations. Foucault is no longer talking about quasi-transcendentals, but about collective political entities. Nonetheless, it remains crucial to recognise that this is life-beyond-individuality, beyond-being, beyond-somatic-bodies – life in relation to which individual organisms are but a passing moment.

This chapter has developed a specific conception of the biological life that enters political history in the nineteenth century and set out (something of) the formal difference between the politics of managing, maximising and securing that life and disciplinary anatomo-politics. The central claim has been to insist that 'biology' in Foucault refers to a historically produced and specific formation of embodiment, to trans-organic population embodiment and *its* vitality. Not to living bodies or the somatic conceived in a general or ahistorical sense. Discipline and biopolitics address very different types of embodiment. They are not simply different scales of operation of a singular politics of the body; rather, they are operationalised in fundamentally different regimes of knowledge with different realities and different rationalities. They are coextensive and often cooperative, but they apprehend wholly differ-ent ontologies, operationalising wholly different forms of analysis and assessment. The next chapter will utilise this conception of life and trans-organic embodiment to illuminate the production and character of specifically political embodiment. Bringing the issues of politics and political *values* into focus, Chapter 2 will pave the way for the remain-ing chapters, which explore biopolitical values, the applicability of the concept of biopolitics in the analysis of contemporary politics, and the relationship between biopolitics, positive critique and feminist political discourse.

2

Incorporation: Foucault on the Co-Constitution of Modern Embodiment, Experience and Politics

This chapter will draw on Foucault's genealogy of sexuality and the above account of biopolitics as the politics of trans-organic population life (Chapter 1), to explain the link between the production of population life and that of vitalist immanent biopolitical values. Vitalist ontologies and ethics, indeed the very existence of *life* as something that could be the ground of political or epistemological evaluation or an end in itself, are conditioned upon the material historical production of trans-organic biological embodiment. Conversely, the production of biological embodiment is a process of *vitalisation* – not of objectification, physicalisation, or reduction. Biological embodiment vitalises and produces vitality – creative evolutionary life – as experience and values, not simply as objects of power.

This historical link – the co-constitution of modern, specifically biological, embodiment, experience and the values of life (of biopolitics) is crucial to the potential of Foucault's account of biopolitics for modern and contemporary political economy. On the one hand the identification of this link demonstrates the *positivity* of bio-mentality and of the production of biological, biopolitical subjects (which is not to say 'goodness' but is to say expansion of force and of limit-experience). Biological embodiment *increases the forces of people's bodies;* it is a process of investment, which goes a considerable way towards explaining the *appeal* or 'hold' of biological knowledge and power. On the other, the identification of this link suggests that life is dependent upon the accumulations and connections of capacities that are constituted through technologies of biopolitical embodiment (such as sexuality). It does, as such, radically problematise any conception of biopolitics that rests upon either an actual or virtual separation between 'life itself' and the trans-organic embodiments, the populations, with which the history of biopolitics is bound, calling into question the idea that the contemporary (or any) period might be characterised by a biopolitics without populations as well as those articulations of vitalist ethics or politics that strive to free life from biopolitical power.

The link between the production of biopolitical embodiment and life as value is implicit within Foucault's analysis, and is explicitly observed with respect to a number of empirical phenomena, most extensively that of sexuality. It will, however, be necessary to elaborate upon that link in this chapter, spelling out an affective, experiential analytics of embodiment that is less than explicit in Foucault's writings on biopolitics.

Whilst literature on Foucault's genealogies of sexuality tends to focus on the politics and production of sex and sexual identity, I want to stress that, for Foucault, the history of modern sexuality is also the history of the trans-organic vital bodies of modern politics in a more general sense – of the tightly bound affective nuclear family, the bourgeoisie, the proletariat, the (specifically modern) race and the universalised nation. In emphasising the strand in Foucault's history of sexuality that leads to modern European racism and nationalism I am following the lead of Ann Stoler's seminal *Race and the Education of Desire*. Stoler draws out Foucault's description of the interplay between productions of sexuality and of race, and develops this analysis in the context of productions of sexuality in the Dutch East Indies. My own analysis will be more abstract pertaining to the production of modern political embodiment *per se*, and moving further away again from the politics of sexual identity and practice. For the purposes of the current study Foucault's genealogies of modern sexuality are interesting insofar as they are critical analyses of modern experience, not as accounts of the politics of sex. Whereas Stoler draws primarily on two of Foucault's key texts on sexuality and race (Foucault, 1978; 2003b) and critically builds upon his analysis through research in colonial history, I will situate those texts within the context of Foucault's oeuvre, render explicit some of the experiential analytics that are at work in his theorising, and demonstrate that modern political (trans-organic) embodiment is a concern that runs through great swathes of Foucault's work. Stoler brilliantly elaborates upon what the history of sexuality tells us about colonial practices of domination in the Dutch East Indies. I want to render explicit Foucault's claims concerning the embodiment of European modernity and to raise the role of positivity, affirmation, affect and connectivity – experience – in the animation and appeal of self-productive nationalism, class-supremacism and racism. As such, I want to render (more) explicit Foucault's contribution to the *positive critique* of modern bourgeois nationalism and the rhetorical self-affirmation of the securitising liberal state. In Foucault's analysis nationalism does not appear as 'false propaganda' or 'ideology' but as real processes producing densities of affective connections and capacities, densities of experience; the nation as a real singular-plural embodiment, a population, in which vital forces of biopolitical subjects are invested.

Foucault is sometimes accused of reducing the affective and embodied to the rules of formation of linguistic discourse (Thrift, 2007). I would argue, to the contrary, that Foucault's analysis of discourse, rationality and rhetoric is wholly intertwined with an analytics of affect and embodiment – of passions that grip you for no reason, have no origins, and are mobile without being directed towards a given point (Foucault, 1996:313; see Robinson, 2003). In particular, Foucault's analyses are informed by a kind of Spinozist sense of embodiment, which is to say a sense of embodiment as constellations of capacities and forces that are in themselves affective and expressive. According to Spinoza, 'affects are the states of awareness of bodily transitions in activity and passivity – transitions in bodily power and intensity' (Gatens & Lloyd, 1999:52). Increases in power and intensity are experienced as joy whilst decreases in powers and intensity are experienced as sadness. The powers of bodies – embodiment – is thus a plane of meaningfulness and value in and of itself, before and autonomously from systems of signification. Whilst the extent to which Foucault can rightly be regarded as a Spinozist is contested (see Juniper and Jose, 2008; Macherey, 1992) this basic analytics of the experiential, affective reality of bodily forces seems to me undoubtedly present in Foucault's genealogies. Foucault often uses the term 'investment' when referring to the relationship between bodies and powers. In using this term he is, I think, not only alluding to economic logistics of productivity but also to the psychologist's use of the term – to *affective* investment. Transforming the forces of bodies is an immediately affective event. Attachments are formed between knowledges, authorities, bodies and subjects in these events of investing power. The emergence of biological knowledge was an immensely significant event in the structure or formation of experience and embodiment, transforming and multiplying locations and pathways of affect and shaping the kinds of values, rhetorics and epistemic mechanisms that could be deployed and generated. In particular the organisation of experience through biological knowledge is a condition of the immanent, quasi-transcendental, vitalist values that are deployed in biopolitical rationality.

Trans-organic embodiment, vitalist values and the temporality of experience: an abstract overview

It is widely maintained that, for better or for worse, the analytical tools and evidence production of evolutionary biology undermined the theological structuring of values, meaning and authority of the 'Western' Judeo-Christian traditions (e.g. Grosz, 2004; Weikart, 2004). The theory of natural selection provided a rational and worldly explanation of the creation and ordering of the living world as finite, even mechanistic, processes. It described how complex, normative, creative processes and 'progress' could have been achieved *within* the world, as the result of random accident and chance variation, without the intervention of either divine or rational judgement and design. With this, the theory either 'tragically undermined' or 'liberated us from' the conceptions of genesis, morality and authority of Judeo-Christian faith and tradition – conceptions that hinged upon the primacy of a divine eternal transcendent domain and power. My concern here is not to engage directly in these metaphysical and scientific debates, but rather to explore the structures of experience that form their context. I am interested in the 'positive unconscious' of biopolitical rationality and rhetoric. I want to emphasise the impact of evolutionary biology upon the rules of formation of *experience* and *embodiment* and the importance that this itself has had with respect to the production of immanent vitalist biopolitical values. Biological knowledge is not simply an event in the history of ideas, it is also an event in the history of experience and embodiment. The latter has been immensely significant in itself with respect to the production and success of values, rhetoric and political discourse within modernity.

My central claim is as follows: Evolutionary biology participated in an expansion and multiplication of experienced embodiment for people who were captured within (or captivated by) biopolitical discourse. Hereditary flows of evolution, degeneracy and reproduction carried health, affliction and propriety into future generations and into the present future life of other people. Bodies spill over into each other in biological perception, in a way that is not true of bodies as apprehended in classical knowledge. This extension of embodiment (beyond the somatic) radically intensified the sense of the importance and meaningfulness of the present moment, the immediate future and of the private. Vitalist biopolitical values make sense of and express this new immense significance of the present and the private. Evolutionary biology gave form to a trans-organic embodiment and this transformed the present and the ubiquitous into the immensely significant *potential*; potential to impact upon the future of countless other lives that constitute the present and future of the community, the nation, the race or the species.

One's present mundane and most private actions really *matter* in the context of genetic, trans-organic, embodiment – especially as that embodiment came to be formulated in legal psychiatry and theories of degeneracy. They matter not only (as they do in the Christian pastorate) with respect to the personal concerns of eternal salvation and moral correctness. Henceforth they matter because they *are* the matter of the health, success and potentiality of a countless, potentially infinite, series of other people's bodies. The previously aristocratic privilege of potential historic significance could become the property of every body – or at least of every body that partakes (potentially) in sexual reproduction or the administration of care for the population. Potential historic significance is no longer the preserve of the very few, made manifest in public spectacle. It is now the common property of a general populace and is made manifest in the (previously) private domains of corporeality and care.

This expansion, intensification and temporalisation of present time is an event in the formation of experience and embodiment that underscores a widespread, generalised, production of immanent values – a production of which the scientific and theological debates over the metaphysics and ethics of evolution is but one part. Since the nineteenth century there has been an immense profusion of political rhetoric, ethical discourse and aesthetics that centre upon the valorisation of the immanent potential of bodies: evolution, vitality, creativity, *life*. The formation of embodiment as population life is the experiential structure or condition of the production of those vitalist values.

Foucault's analysis suggests that the lynchpin of the production of vitalist values in the context of biopolitics is the establishment of a connection between the intimate forces of people's bodies, and the vitality of the trans-organic life of population. This connection is addressed at length in Foucault's writings on sexuality and sex.[1] Vitalist and biopolitical values are sometimes associated with modern individualism – the affirmation of the creative and emancipated individual. Foucault's account of the transformation of embodiment in the formation of biopolitics suggests, however, that we should understand that very individualism as a part of a broader collectivism; the biological and biopolitical formations of embodiment render individual bodies already a part of collective, trans-organic, extra-individual life, such that the pursuit of individual creation, emancipation, health or normality is *already* a part of the pursuit of collective life and of connectivity, sociality, or escape from finite singularity.

Having established the broad outline of my argument with respect to the relationship between trans-organic embodiment, the temporality of the present and the production of biopolitical values, we will now move on to a more detailed consideration of the production of that embodiment in Foucault's writings on sexuality. We will begin with a brief overview of Foucault's work on sexuality before moving on to a series of themes through which Foucault explores the history of extending, biological, modern political embodiment in the context of sexuality. Sexuality is a series of modern inventions, including: theories of sexual reproduction, heredity, degeneracy and evolution; the sexualisation of the family and of childhood; the development of class distinction through sexuality; and many other things besides. These inventions created new formations of embodiment and experience, resulting in profound transformations in the possibilities and ordering of values and political discourse.

Foucault on sexuality and subjectification

The key texts in which Foucault explores biopolitical sexuality are *The History of Sexuality: 1* (1978) and the *Abnormal* lecture series of 1974–75 (2003a), the latter containing much of the empirical substance behind the arguments in the former. Here Foucault addresses sexuality in its biopolitical aspect, with the concept of biopolitics appearing for the first time in the conclusion of *The History of Sexuality: 1*. Foucault is interested, here, in the great significance that sexuality and sex have come to bear in the context of European modernity, both for individuals searching for meaning in their soul or identity, and for political and scientific research seeking to understand and govern population life. Ours, he suggests, is 'a society of sex', a society in which sex is what's worth killing and dying for (Foucault, 1978:156). A part of the explanation for the importance that has come to be attached to sex is that sexuality produces the collective bodies that define the politics and governance of European modernity – the bourgeoisie, the nation, the proletariat, the race and civil society. These collective bodies come into clearer focus in Foucault's subsequent lecture series (especially 2003b; also 2007, 2008a). In those lecture series Foucault also develops the theme of arts of government and governmentality (the conduct of conduct), drawing attention to the governance and securitisation of society through the production of autonomous individuality. Foucault then returns to the topic of sexuality in *The History of Sexuality 2: The Use of Pleasure* (1985a). Here Foucault sets aside the biopolitical context to focus upon an alternative production of individuality – the Greek aesthetics of existence – in which sexuality was a significant focus for work upon the limits of the self.

The History of Sexuality: 1 is addressed to the huge importance of sexuality within modern European culture, illuminating and making strange the profusion of discourses concerning sex and sexuality since the eighteenth century as well as the role of the constructed element – sex – in the production of identity. 'At the level of discourses and their domains' there has, Foucault contends, been 'a steady proliferation of discourses concerned with sex...a discursive ferment...an institutional incitement to speak about it, and agencies of power to hear it spoken about, and to cause it to speak through explicit articulation and endlessly accumulated detail' (1978: 7). A crucial element in this discursive explosion was the installation of the idea of sex as a specific, hidden and productive element in our being (an element that masks the role of power in its own production, and thus what gives power its power) (Ibid.: 150–7).[2] The importance of sexuality and sex is manifest not only through the immense institutional and political interest in the regulation and investigation of sex, it is also manifest through the purported role of sex as something like the opposite to power, institutions, the State and the corruption that they wield. Sex and sexuality are called upon to grant people access to the very truth of their being and to their place in society and history – to be a measure of authenticity and autonomy or liberation:

> It is through sex – in fact, an imaginary point determined by the deployment of sexuality – that each individual has to pass in order to have access to his own intelligibility..., to the whole of his body..., to his identity...– we have arrived at the point where we expect our intelligibility to come from what was for many centuries thought of as madness; the plenitude of our body from what was long considered its stigma and likened to a wound; our identity from what was perceived as an obscure and nameless urge. (1978:155–6)

Foucault sets out his first account of biopolitics in the conclusion to *The History of Sexuality: 1,* as a contextual explanation for the significance that sex and sexuality claim in modern European society. As we saw in Chapter 1, Foucault maintains that modern power, orientated upon the production and protection of life, developed along two distinct poles; the disciplinary anatomo-politics of the human body and the biopolitics of the population, or of life. Sex, Foucault argues, comes to play such a significant role in modern discourse and values because it is situated *at the intersection* of these two poles of power. Sex is a means of access both to population life and to the individual body:

> On the one hand [sex] was tied to the disciplines of the body: the harnessing, intensification, and distribution of forces, the adjustment and economy of energies. On the other hand, it was applied to the regulation of populations, through all the far-reaching effects of its activity. It fitted in both categories at once, giving rise to infinitesimal surveillances, permanent controls, extremely meticulous orderings of space, indeterminate medical or psychological examinations, to an entire micro-power concerned with the body. But it gave rise as well to comprehensive measures, statistical assessments, and interventions aimed at the entire social body or at groups taken as a whole. Sex was a means of access both to the life of the body and the life of the species. (1978:145–6)

It is not simply that sex is interesting from the perspective of two different regimes of power and is thus doubly interesting or significant. Rather, it gains an immense significance because it forms *the point of contact between* these two regimes of power and the two forms of embodiment to which they correspond. Sex is a means of access to *both* the life of the individual somatic body and the life of the trans-organic population at the same time. Sexuality is one of the crucial mechanisms that invest somatic bodies in trans-organic population life. It is crucial to the very production of trans-organic embodiment, constituting flows between the limited bodies of organisms in populations. It is this role as link and intersection – transcending limits – that makes sex such a potent rhetorical, ethical and affective tool.

For Foucault, if sex is a means of access to one's intelligibility, to the whole of one's embodiment, and to one's identity, it is because one's sex constitutes one's connectedness to the trans-organic, historic, forces of life – forces which unfold in the hidden depths of passing time. One passes through sex to gain access: to one's own intelligibility, because 'it is both the hidden aspect and the generative principle of meaning'; to the whole of one's body, because 'it is a real and threatened part of it while symbolically constituting the whole'; and to one's identity, because 'it joins the force of a drive to the singularity of a history' (1978:155–6). Sex is granting individual somatic bodies access to meaningfulness, collective embodiment and to history.

The significance of sexuality and sex for power and for the practices of self, is tied to the role of sex in producing modern political biological community: class, race and nation. As we will elaborate upon below, this production is about the empowerment of bodies to affect and be affected by other bodies – to be of political and historic significance. There is a parallel between this and the apparently individualised ethics of sex amongst the free men of ancient Greece – the subject of Foucault's *History of Sexuality: 2*.

Work upon sexuality in the context of the free men of ancient Greece was, according to Foucault, constructed as an individualised creative ethics – a work upon the self, guided by the equivalent of self-help books and philosophical discourse, but executed as an autonomous, practical and creative art of individual living. Without wishing to digress into a discussion of non-biopolitical Greek sexuality, it is worth noting that here too, even in the context of a seemingly radically individualised ethics of existence and self, work upon sexuality is rendered significant *because* of its importance for the production and protection of the political community of the *polis* (the political community), the city and home. The arts of sexual life – undertaken as a kind of practical training by the free men of Greece – pertained to the development of an effective and informed self-mastery that would enable these men to rule as democratic and just autonomous citizens. The idea of autonomy was crucial to the Greek conception of democratic politics. One could participate in the polis only to the extent to which one was free. To be free, however, was almost the opposite of being oneself. To be free and capable of participating in the democratic polis meant being free from all self-interest, including concern with one's own survival or body. (Hence the absolutely integral nature of slavery and the political subjection of women to the Greek political system – the subjection and separation of those that *do* care for self-survival and embodied life [the domains of necessity in which men rule as 'just' masters not democratic citizens] without which the separate democratic domain of freedom and action could not possibly exist.) In establishing a practical and strict control over one's sexual urges free men could foster and prove their 'freedom', their capacity to be stronger than themselves, than their own needs, and thus adequate to the task of politics (Foucault, 1985a:139; 250). The arts of sexuality also pertain to the establishment of a just relationship with one's wife, and thus of justice in the home or *oikos* (Ibid.:178) and to knowing how to master one's own pleasure without infringing upon the freedom of other free men in the making – the boy that is the object of erotic love (Ibid.:252). The sexual arts of life in ancient Greece are, as such, invested in the production of the power of free men to act with each other and form the polis, as well as to exercise domination over the city and the home, over slaves and other women. This art of the self is a part of the production of political community, of governed territory and of home.

We will now move on to elaborate upon the role of sexuality in the production of modern trans-organic embodiment, stressing that such production is a matter of establishing flows and relations of power and affect.

Sexuality and the production of trans-organic embodiment: heredity, masturbation, population

Foucault's genealogy of modern sexuality demonstrates how the techniques and discourses addressed to sexual practice and identity have contributed to the production of modern trans-organic embodiment, a production that has, in turn, constituted a new affective plane and historical grid of intelligibility – which is to say a new formation of experience. In this section I will elaborate upon the role of sexuality in the production of trans-organic embodiment (and thus modern experience) through reference to modern theories of reproduction, theories of heredity and degeneracy and the sexualised affective nuclear family. The discussion will draw primarily upon Foucault's texts. The discussion of modern theories of sexuality, however, will also draw upon François Jacob's *The Logic of Life* (1973), which has a strong affinity with Foucault's work, as discussed in the previous chapter. Jacob's work gives a considerably more detailed account of modern theories of reproduction and its predecessors than does Foucault's, helping to illuminate the role of biological knowledge in the transformation of understandings of creativity and thus of experience.

Modern sexuality, a historical constellation of technologies of power/ knowledge, contributes to the augmentation of present embodiment and the production of trans-organic, limit-transcending, life. Theories of sexual reproduction, which were invented in the context of modern biology, intensified the relationships between bodies. In particular this theory attributes bodies with the capacity to create new organisms, which is in radical contrast to the theories of preformism that were dominant in the seventeenth and eighteenth centuries, which insisted that organisms must have been created, in germ form, at the time of divine creation (Jacob, 1973:57, 177). Medical discourses on child sexuality and, later, psychiatric theories of degeneracy, rendered *practice* capable of penetrating these creative reproductive processes, such that practices directed at managing health could impact upon the potentially infinite future of hereditary population life (Foucault, 2003b:315). The regulation and discipline of other people's sexuality – especially that of children – would enfold bodies into each other, generating trans-organic flows of influence and sites of potentiality (Foucault, 2003b:248). In each of these respects sexuality participates in the formulation of somatic bodies as finite *moments*, limits and participants within the potentially infinite process of life, incorporating somatic bodies into trans-organic embodiment, the population, life which perpetually transcends its own limits. This trans-organic embodiment is not *a body*, in the sense of being either a physical mass or an organism (organised being). What might be individuated as a trans-organic embodiment, *a* population, is the constellation of *connected capacities*, connected vitality. The theory of sexual reproduction, which is developed in modern biological knowledge (bio-mentality), conceives of organisms' capacities as being connected: vertically connected across generations by heredity; and horizontally connected to other present organisms through the potentiality of a common future. It radically politicises somatic bodies, making them the matter of each other, locating the creation of life within the body of organisms and transforming the historical grid of influence, intelligibility and significance. The creation of life is no longer a thread stretching throughout creation, linking present individuals to a transcendent realm and divinely ordained order. The creation of life binds people to other people, to their practices and (previously) private concerns. The creative present of modern experience is the sexualised present of modern biological knowledge.

The theory of reproduction

Jacob recounts a detailed picture of modern theories of generation in terms of sexual reproduction, situating its emergence historically, and contrasting it with earlier theories of generation: with spontaneous generation, in which divine creation intervened directly in the creation of life as Descartes believed; and with preformism, where all beings were created at the same time, the time of divine creation, and stored up in miniature to unfurl in the present. These different theories of generation do, then, pertain to very different formations of dimensionality, formations of experience.

Reproduction, Jacob suggests, is the central concept of modern biology. 'For modern biology', he writes, 'the special character of living things resides in their ability to retain and transmit past experience'. For biology, an individual organism is 'merely a transition, a stage between what was and what will be. *Reproduction* represents both the beginning and the end, the *cause* and the *aim*' (Jacob, 1973:2, italics added). The term 'reproduction' was first used to refer to the generation of living organisms by Buffon in 1748 (Ibid.:72). It did not come to refer to the concept of sexual reproduction – in which the actual creation of organisms comes to be located in the sexual act and the parents' bodies – until the nineteenth century.

Throughout the seventeenth and eighteenth centuries the dominant scientific view on the generation of organisms was 'preformism'. Preformism rejects the earlier thesis of creation through direct divine intervention in the world – spontaneous generation – and proposes that organisms develop from a seed or germ. The theory did not, however, attribute the creation of the germ to the sexual act. Rather preformism maintained that germs were microscopic versions of the adult organism which had already existed as a completed whole within one of the parent's bodies. Debates raged over whether it was the mother or the father in which the germ resides. As Jacob puts it:

> [The germ] is like a scale model with all the parts, pieces and details already in position. The complete, although inert, body of the future being lies already waiting in the germ. Fertilization only activates it and starts it growing. Only then can the germ develop, expand in all directions and acquire its final size, like those Japanese paper flowers which, placed in water, unwind, unfold and assume their final shape. (1973:57)

The theory of preformism implied that of 'pre-existence'. Although preformism could account for the generation of a given organism, it only deferred the question of creation – for these perfectly structured forms *must* have come from somewhere. To the classical mind this 'somewhere' meant a divine intelligence. In the classical mindset it made the most sense to attribute the genesis of the germ to the time of creation. All the germs of all the organisms past, present or future were 'pre-existent', divinely created at the dawn of mortal time. Either the female or the male germ must, as such, contain 'the germs of all its descendents, nested into one another like Russian dolls' (Jacob, 1973:61). In the theories of preformism and pre-existence, creativity and the generation of potential are firmly cordoned off from mortal activity and finite lives. This will be radically transformed with the development of knowledge of sexual reproduction. The ideas of preformism and pre-existence began to be undermined in the eighteenth century, when Buffon, for example, demonstrated – mathematically – that the size of the germ within the germ ... would be preposterously small, and Maupertuis demonstrated that offspring regularly inherit abnormal features from *both* mothers and fathers, undermining the idea that the germ could have preceded sexual union (Ibid.:67, 70). In the nineteenth century, preformism and pre-existence were displaced, in scientific circles, by the theory of reproduction.

The theory of sexual reproduction gained an increasing hold in the nineteenth century with the development of modern biology, which looked beneath the surface of organisms, introducing a new depth into analysis (as set out in Chapter 1). The theory of reproduction has undergone two major reconfigurations since its original formulation at the beginning of the nineteenth century, moving from the level of the organism to that of the gene, and from the level of the gene to that of nucleic acid (DNA). The first stage, however, constituted the most fundamental reconfiguration of the relationship between bodies, time, divinity and processes of creation. Jacob writes:

> As long as living organisms were perceived as combinations of visible structures, preformation provided the simplest explanation for the persistence of those structures through succeeding generations. The linear continuity of the living world in space and time required a continuity of form through the actual process of generation ... Filiation, therefore, had to have the same inertia as the whole system. (1973:74)

With the emergence of the science of organisation, life and the concept of reproduction (the emergence of modern biology), Jacob continues, 'the very nature of empirical knowledge was gradually transformed' (Ibid.). Perhaps most crucially from the point of view of the organisation of experience, the very cause of the existence of living bodies would now, for the first time, be located within living bodies themselves. The theory of reproduction conceives of bodies as actively connected, causing, creating, each other.

At first the concept of reproduction meant that 'a new relationship appeared between beings, linking individuals vertically through generations' (Jacob, 1973:141). With the theory of natural selection and evolution, which developed in the middle of the nineteenth century, that connectedness extended to different species. After Darwin, and the theory of natural selection, reproduction came to be 'the main factor operating in the living world [according to biologists], the source of both stability and variation'; it was the meeting point of determinism and contingency, the cause of creativity and stability (Jacob, 1973:177). This incorporation of the process of creation into those of mortal bodies fundamentally transfigured the relationship between knowledge, the world, divinity and observation. The mysterious realm of divine determination is destroyed, or rather enfolded into the processes and physicality of mortal life. With this the very act of creation was rendered scientifically observable. Jacob writes:

> If no intention was responsible for the appearance of new forms, then their success or failure in the 'struggle for existence' depended only on physical factors, that is to say, variable parameters. No sphere of biology remained inaccessible. Even reproduction could become the object of investigation. (Ibid.:177)

The theory of reproduction did, then, reconfigure somatic bodies as causally connected – the location of each other's creation and determination. That reconfiguration of embodiment meant a radical transformation in the organisation of experience, the production of values and truths, and the role of divinity (or the outside of finite existence) in the finite observable world. Sexual reproduction brought divine and mysterious processes of creation into the material, finite and (potentially) knowable processes of living bodies – grasping the connected capacities of those bodies, their trans-vital, trans-organic embodiment. The dimensionality of the world, the dimensions that needed to be present for the world to be possible, was radically transformed, with acts of creation and determination enfolded into the finite and mortal domains of history.

Heredity and degeneracy

Of course theories of heredity and sexual reproduction were not only developed and popularised by biologists. In *The History of Sexuality 1* (1978) and *Abnormal* (2003a) Foucault emphasises the theories of heredity that were developed in medico-legal and educational practice. These theories emerged at the turn of the nineteenth century. The latter half of that century saw two major developments in the technologies of sex that were both underscored by the theory of degeneracy: the medicine of perversions and the programmes of eugenics. The theories of heredity that Foucault writes about, such as degeneracy, are rather murkier than the modern biological theory of reproduction, as Jacob describes it. The most significant difference is that nineteenth-century medical and legal conceptions of heredity conceived of *learnt behaviours* associated with health and especially sexuality as biologically heritable and as of (immense) evolutionary significance.[3] Sexual perversions were understood to be deeply invested within bodies and to be a determinate of the general health of bodies – not only of perverted individuals but also of their children, and their children's children, and – given the looseness of the concept of sexual inheritance – perhaps of any body with which it came into too close contact. The theory of reproduction incorporated creativity into present mortal bodies. The theory of degeneracy treated that capacity as contingent, as something that could be transformed or infected by practice.

Bodies, especially the sexual practices of bodies, were being *made responsible* for the health, vitality and life of the species itself (Foucault, 1978:188; 2000j). In the mid-nineteenth century:

> the analysis of heredity was placing sex (sexual relations, venereal diseases, matrimonial alliances, perversions) in a position of 'biological responsibility' with regard to the species: not only could sex be affected by its own diseases, it could also, if it was not controlled, transmit diseases or create others that would afflict future generations. (1978:118)

In situating the creation of life within the bodies of organisms – and especially within the sexual practices of organisms – modern theories of heredity had made bodies and their sex responsible for the future life of the species. This opened bodies up to the intervention of power. Foucault argues that with these theories of heredity it became possible to hold people legally responsible, not simply for what they had *done* but for what they *are* (Foucault, 2000j). Psychiatry developed as a science not of illness but of *abnormality* and degeneracy, addressed to the danger that perverted bodies present to the collective body. An entire medico-legal apparatus developed around bodies and their sexuality, intervening in sexual practice for the sake of a general protection of society and the race (Foucault, 1978:119; 2003b:317). In Foucault's analysis the force of sexuality as a discourse and apparatus is vested within its capacity to make matter matter. And what mattered on an evolutionary scale, given the murkiness of heredity as conceived in medico-legal and educational discourse, included practices that could be learnt, controlled and eliminated. The medico-legal – essentially psychiatric – conceptions of heredity were, in effect, more *constructivist* than the theories of reproduction of biological science. They placed immense significance upon the capacities of society, and especially parents, to construct the future health of their children.

What Foucault makes rather less explicit is that this 'biological responsibilisation' of bodies and their sex is also a kind of empowerment or augmentation of those bodies. Making biology responsible did not only open bodies up to power; *it opened power, the future and historical significance up to bodies*. Bodies – their inner being and their most private practices – gain an enormous significance and capacity in the context of biological heredity. By tracing and making manifest connections between bodies, behaviours and capacities, the great medico-legal and educational apparatus of sexuality made bodies potentially infinitely significant. There is a massive *positivity* in the development of sexuality, which is to say a massive generation of force and capacity in present bodies.

In *The History of Sexuality: 1* Foucault attempts to make this positivity of the technologies of sexuality evident by pointing out that sexuality was developed first by the bourgeoisie *for the bourgeoisie* (1978:122–7). Technologies of sexuality were developed, in the first instance, not as mechanisms of domination, control or negation, but as a series of 'techniques for maximising life' – techniques that would enable the bourgeoisie to feel that they had the right to govern, to feel themselves distinct, special and empowered. Foucault writes:

> What was formed was a political ordering of life not through an enslavement of others, but through an affirmation of self...[the bourgeoisie] provided itself with a body to be cared for, protected, cultivated, and preserved from the many dangers and contacts, to be isolated from others so that it would retain its differential value. (1978:123)

A key domain in which to care for the sexualised body, and thus for the very life of the population or bourgeoisie, was the family. A particular concern was the sexuality of children.

Masturbation and the family body

Foucault devotes one of the *Abnormal* lectures to the discursive explosion concerned with child masturbation, which took off in the latter half of the eighteenth century in Europe and did not begin to subside until the end of the nineteenth (Foucault, 2003a:231–58). The issue of masturbation is particularly pertinent for the present discussion because it highlights the ways in which regulative sexuality (including medical expertise concerning sexuality) enfolded parents' bodies into and around the life and potential of their children – orientating the family towards potentiality and youth.

Childhood masturbation was said to be the cause of terrible, even lethal, illness and decline. As theories of heredity and degeneracy took hold, such illnesses were thought capable of inflicting future generations. Masturbation becomes, in these discourses, a matter of genuinely vital concern – not only for individuals but for the very species. Parents are implored to protect their children – and thus society – from this terrible vice. Foucault argues that this bourgeois obsession with child masturbation was a part of the process by which the modern, nuclear, closely bound, affective family was formed – the family that was formed as a singular-plural body and plugged into medical and State power. In the nineteenth-century campaign against masturbation we see the regulation of childhood sexuality giving body and historicity to the family (and thus to the population of which the family is a part). The campaign against masturbation can, as such, be seen as a part of the production of modern embodiment and present-centred, historicised, experience.

Foucault refers to a discursive explosion around masturbation. This included books that were produced for parents, as well as for adolescents themselves, on the avoidance of masturbation. It included clinics that advertised their services for curing children of this vice. At one time there was even a wax museum in Marseille, in which children could be introduced to a theatre of wax models depicting all the terrible aliments that would be wreaked upon the bodies of masturbators (Foucault, 2003a:235). Texts and spectacles aimed at children warned them of the hideous deformities, diseases and eventual demise that would be the result of their masturbating. Texts aimed at parents informed them that *they* were the ones who were responsible for this vice and implored them to do their duty and keep a constant watch on their children. Foucault quotes from Rozier's 1806 *Lettres médical et morales*. Rozier wrote:

> Parents…who, through a blameworthy lack of concern, allow their children to fall into a vice that will lead to their ruin expose themselves to the risk of one day hearing the cry of despair from a child who is dying while committing a final offence: 'Woe to who has caused my ruin!' (quoted in Foucault, 2003a:245)

The discourse surrounding masturbation places a radical and direct responsibility onto parents for the health, not only of their own children but of their children's children and even for the vitality and security of society or the race. Parents were told that they were responsible for the sexuality of their children and were called upon to keep adolescent bodies under perpetual surveillance. Foucault argues that this constant surveillance and responsibilisation amounts to *a reorganisation of the physics of the family*. There is an application of the parent's body to that of the child, an enfolding that incorporates the potentiality of the child into that of the parents. This enfolding, applying, connecting and sexualising of family members' bodies resulted in the production of the nuclear family, which is to say the family *as* body. Foucault writes:

> Beneath these puerilities *[these injunctions about how to prevent children from masturbating]* there is … a very important theme. This is the instruction for the direct, immediate, and constant application of the parents' bodies to the bodies of their children. Intermediaries disappear, but positively this means that from now on children's bodies will have to be watched over by the parents' bodies in a sort of physical clinch. There is extreme closeness, contact, almost mixing; the urgent folding of the parents' bodies over their children's bodies; the insistent obligation of the gaze, of presence, contiguity, and touch … The parent's body envelops the child's and at this point the central objective of the manoeuvre or crusade [against masturbation] is revealed: the constitution of a new family body. (Foucault, 2003a:247–8)

Discourses and techniques surrounding the regulation of childhood sexuality are central to the formation of the modern nuclear family. That family is 'a sort of restricted, close-knit, substantial, compact, corporeal, and affective family core' (Foucault, 2003a:248). This is an embodied family – a corporeality – composed of affective investments and flows of power. As will be explained below, Foucault argues that this family was not a closed, singular, private body but rather a component of the great trans-organic body, the population.

Foucault describes the regularisation of the family as the reorganisation of the physics of the family. This is not a specific phenomenon; rather, the reorganisation of the physics of the family is a part of the reorganisation of physics *per se*, of dimensionality, temporality, visibility and light. It is an aspect of the reorganisation of modern experience: a production of trans-organic embodiment (with the capacities of bodies enfolded into each other) which constitutes the present as expanded, creative, historicised and profoundly political.

Family body as component of population

In *Abnormal* (2003a), Foucault develops the idea that the bourgeois concern with sexuality in the eighteenth and nineteenth centuries is a part of the production of collective embodiment. As we have seen, Foucault claims that the concern with such activities as childhood masturbation produces the family *as body* through the sexualisation and responsibilisation of bodies. Further, Foucault claims that this sexualised family body is also a medicalised body, a body invested with medical knowledge and power. The power that flows in the family is not a private power. Rather, Foucault argues, this sexualised nuclear family is plugged in to a whole external medical power/knowledge. The corporeal, dense, affective, sexualised family is 'the medicalised family' (2003a:250). Not only has the family become a singular-plural body itself, it is also a part of the collective body that is regulated through medico-legal power; families are components of populations.

Foucault develops this theme in the *Security, Territory, Population* lectures (2007:104–5). From the mid-eighteenth century onwards, he argues, the family begins to be conceived of as a *component part* of the population and as an *instrument* for the governance of population life. Whereas the family used to act as a *model* or *metaphor* for government, it became, with the emergence of the trans-organic population, instead an actual segment of the population that is to be governed. The family is privileged in the government of population life, but it is not an *end* of biopolitics in itself. Rather, the family is privileged as the point of governmental access to the sexuality of bodies; 'when one wants to obtain something from the population concerning sexual behaviour, demography, the birth rate, or consumption, then one has to utilise the family' (Ibid.:105). Since the mid-eighteenth century the (emergently nuclear) family has, Foucault claims, existed in a fundamentally instrumental relationship to the population. The family is an instrument of biopolitics, a means of access to and regulation of population life.

It is worth emphasising that this investment of biopolitical power in the sexualised bodies via the family runs in both directions – it is an *investment* of power, not an exercise in domination. The family is privileged from the perspective of the regulation of the population because it is through work upon the family that one gains access to the sexuality of bodies. Conversely, the family is the point of access to population life and power on the part of the sexualised somatic bodies and their practice. Through sexual reproduction and the creation of family, parents created the life of the new generation. By becoming 'fully and truly parents' (2003a:249), by protecting the little masturbators from their personal perversions, parents could exercise power of historical significance, protecting not only their own children but the population itself from the evils of degeneration. By preventing one's child from masturbating one might be safeguarding the future of civilisation! The newly affective, sexualised, responsibilised roles of family life do not only subordinate family members to the demands of the population; they grant the practice of parenting *public* significance, as well as a legitimate claim upon medical and governmental power. Once parenting becomes such an important task, parents are surely entitled to the assistance of doctors, governments and scientists: knowledge and power.

Foucault claims that this nuclear family (as body, as instrument of population) began to emerge in the eighteenth century, and was the matter of intensive work – of discursive explosions and technological innovations – throughout the nineteenth. By the twentieth century, the population, to which the family was subordinated as instrument, was the national body of the nation state. In the eighteenth century, however, the population in question was a much smaller collective; it was the 'class body' of the bourgeoisie. The following section will explore Foucault's writings on the production of a bourgeois class body, its relationship to bourgeois domination and to the nation and nationalism, arguing that nationalism can be understood, historically, as the generalisation of bourgeois class-racism and self-affirmation.

The key point to take forward from this section is that sexuality – the sexualisation of the family and the regulation of childhood pleasures, the theory of sexual reproduction and later evolution, and the theories of degeneracy and heredity – transformed the relationship between creativity and ordinary organic life. Sexuality *connected* bodies. But it did not do so through the production of a kind of giant organism or unified mass. Rather, sexuality produced a trans-organic embodiment, into which the potentialities and capacities of different organisms extended, reaching beyond the limits of the organism itself. Those capacities might be the work of parenting, the virility and perversity of desire, or the generation of life itself. Those capacities are many and varied. However they all engender a specifically (biopolitically) modern enfolding of creativity into embodied life. They make the present and (previously) private embodied world immensely significant, responsible, (bio)political.

Biopolitical embodiment: class and nation

The history of embodiment to which Foucault's genealogy of modern sexuality pertains is the history not only of modern sexual identity, but also of specifically political modern embodiments; modern classes, nations and races. Foucault's comments about the bourgeoisie and the nation as body appear in *The History of Sexuality: 1* and *Society Must Be Defended* (1978; 2003b). They are crucial to grasping the *stakes* of biopolitics as Foucault understands it. The history of biopolitics is the history of the production of modern class politics and the nation state. It is also the history of specifically modern races and of the forms of racism that grew out of European colonialism and eugenics. Indeed, as Foucault sees it, the politicised bourgeoisie, proletariat and nation *are*, in effect, races; they are produced as races. Biopolitical discourse and governance constitute social classes and nations as *biologically* distinguished populations or population fragments. Foucault claimed that racism is inscribed as the basic mechanism of power as exercised in modern states (2003b:254).

An aspect of Foucault's genealogies of biopolitics that has been largely overlooked is the role that Foucault attributes to the establishment of bourgeois power in the production of biopolitics, and the original contribution that he makes, through this, to the history of modern classes. Foucault is not generally associated with class history. However both *The History of Sexuality: 1* and *Society Must Be Defended* are fundamentally concerned with the history of class politics. Both discuss modern power and experience in terms of the establishment of bourgeois class hegemony. Proletarian class politics are also central to the concerns of *Society Must Be Defended*. Whereas Marxist analysis focuses on class *conflict*, Foucault describes modern class politics in terms of a productive *self-affirmation* and techniques for maximising life. The nuclear family and modern sexuality developed, according to Foucault, as instruments of bourgeois self-affirmation, maximisation and distinction: *not*, as is argued by Marxist and Freudian critiques of modern sexuality, as an instrument of domination and repression of other classes.

Foucault argues that sexuality enabled the bourgeoisie to produce for itself a class body – a trans-organic embodiment that would serve as the bourgeois alternative to the ancestry and alliances of the aristocracy. Whereas the aristocracy asserted their specialness and right to govern through claims about their blood – about the antiquity of their ancestry, the value of their alliances and the bloody strength of their *past* conquests – the bourgeoisie would base its claims on its progeny, the health of its body and upon its *present capacity* to fulfil the potential of the state as a self-administering nation (Foucault, 1978:106–7, 124; 2003b:223). Sexuality was extended to the whole of the population, becoming integrated with mechanisms of exploitation, at a *later stage* and through a series of struggles and economic emergencies (Foucault, 1978:121–2). Arguably, modern nationalism can be understood as the outcome of this generalisation, such that nationalism is the generalisation of the bourgeois politics and technology of self-affirmation.

Crucially, Foucault's arguments about bourgeois power and embodiment pertain to the character and production of *modern experience* and the politics that this makes possible. Foucault is concerned with the bourgeoisie in the history of modernity, not because he wants to recreate the argument that modern capitalism is about bourgeois class domination and relations of exploitation (an argument which might sensibly be left to Marxist analysis), but because he is suggesting that *in the process* of establishing bourgeois hegemony, in the eighteenth and nineteenth centuries, new technologies of power and knowledge developed that transformed the structure of experience. The grid of historical intelligibility was transfigured. Historical discourse, rights discourse, political strategy and rhetoric came to be focused upon the *present* as the fullest moment of history, rather than upon the past as the source of authority, right and creativity, and upon present potential and life as the object of political struggle.

Producing a bourgeois class body

As we have seen, Foucault's writings on biopolitics are a part of his efforts to establish a *positive* account of power. Nowhere is this attempt more clear than in the passages of *The History of Sexuality: 1* that are addressed to the relationship between the development of sexuality and the establishment of bourgeois hegemony (1978:122–7). Foucault has established a picture of sexuality as a technological web of power; productions, techniques and interventions of expertise in people's bodies, producing sex itself as an effect of power. In agreement with Marxist critiques of modern sexuality, Foucault argues that the development of this great technology of power can be linked to the establishment of bourgeois hegemony. However, Foucault argues that it was not developed in order to enable one class – the bourgeoisie – to take hold of and control the sexuality of another. Sexuality was directed at *bourgeois* bodies long before it was extended to other classes and utilised as a mechanism of institutional control. Sexuality was developed, first, by the bourgeoisie *for* the bourgeoisie. Moreover, it was not developed as a mechanism of asceticism – renouncing or negating pleasure and flesh – but rather as 'an intensification of the body, a problematisation of health and its operational terms: ... [as] techniques for maximising life' (1978:122–3). The development of sexuality was not about the enslavement of others but about an affirmation of self – an affirmation of self that centred upon heredity and sexuality. He writes:

> Let us not picture the bourgeoisie symbolically castrating itself the better to refuse others the right to have sex and make use of it as they please. This class must be seen rather as being occupied, from the mid-eighteenth century on, with creating its own sexuality and forming a specific body based on it, a 'class' body with its health, hygiene, descent, and race: the autosexualization of its body, the incarnation of sex in its body, the endogamy of sex and the body. (Ibid.:124)

The bourgeoisie were creating for themselves a class body, a body that would mark them out as special and as fit to govern, shifting the historical grid of legitimacy from the antiquity of the aristocratic blood line to the health and capacity – potentiality – of the bourgeois, collective, body (Foucault, 1978:124; 2003b:223). In addition to substituting heredity for the aristocratic concern with alliance and ancestry, this production of a class body was a part of an immense vitalisation – potentialisation – of the bourgeoisie. It constituted 'an indefinite extension of strength, vigour, health, and life' (Foucault, 1978:125). It did, as such, produce the new potentiality, the new capacity, that sufficiently engorged the present moment of history to enable political authority to be based upon existent potential, rather than past conquest, right or brilliance (2003b:223). Sexuality and the new bourgeois body effected a transformation in the grid of historical intelligibility.

The new formation of experience

In *Society Must Be Defended* Foucault discusses the establishment of bourgeois hegemony in relation to the formation of the nation state and the establishment of statist universality. His source material is the writings of eighteenth- and nineteenth-century French aristocratic and bourgeois historians – writings through which the aristocrats and bourgeoisie attempted to establish their (competing) claim to the right to rule (their right to the State). In the movement from aristocratic to bourgeois histories we see, according to Foucault, an 'inversion of the temporal axis of demand' (2003b:22):

> The demand [for power] will no longer be articulated in the name of a past right that was established by either a consensus, a victory or an invasion. The demand can now be articulated in terms of a potentiality, a future, a future that is immediate, which is already present in the present... (Ibid.)

The establishment of bourgeois hegemony relates to the development of technologies of sexuality, the production of trans-organic embodiment and to a reorganisation of the grid of historical intelligibility. The incorporation or embodiment of power, represented in the development of sexuality, accompanies a reorientation of political claims and values upon potentiality, the present and the future. We have, in effect, a new formation of experience in which political discourse, rationality and struggle will be formed.

The importance of bourgeois power, in Foucault's analysis of modern politics, pertains less to the issue of 'who rules' – the establishment and perpetuation of bourgeois dominance – than it does to the production of the *economy of experience* in which modern politics takes place. The *process* by which bourgeois class hegemony was established involved the production of technologies of power/knowledge that reorganised the space, time and dimensionality in which politics unfolds, transforming the kinds of values, rhetorics and productions of affect that can animate political discourse and struggle. Biopolitical modern political discourse and struggle takes place within this formation of space, time, dimension and visibility, whether it is the bourgeoisie who are attempting to secure power, or the Bolsheviks, or the Labour movement, or, we could add, the suffragettes, the nationalists or a host of other characters. Foucault's point is not about bourgeois domination or class enmity; it is a point about the character of modern power/knowledge and the experience that it engenders. It is a point about the productive, presentist, self-affirming, life-maximising character of modern biopolitical discourse.

The bourgeoisie produced for themselves a class body and staked their right to power, to the State, upon the vitality and specialness of that body. In the process a whole new spatio-temporal formation of experience – of values, epistemology and affect – is produced. With this, life, the present and potentiality become what is at stake in political struggle. Not just for the bourgeoisie but for politics in general. Life, potential and trans-organic embodiment become the ground of contestation, rhetoric and value in the political world that results from the production of bourgeois power. Not only would the *bourgeoisie* stake their claim to power on their vitality and potentiality; political *challenges to bourgeois rule* would deploy a similar strategy, staking their rights claims on life and potentiality. Foucault writes:

> [A]gainst this power that was still new in the nineteenth century, the forces that resisted relied for support on the very thing it invested, that is, on life and man as a living being. Since the [nineteenth] century, the great struggles that have challenged the general system of power were not guided by the belief in a return of former rights, or by the age-old dream of a cycle of time or a Golden Age. One no longer aspired toward the coming of the emperor of the poor, or the kingdom of the latter days, or even of the restoration of our imagined ancestral rights; what was demanded and what served as an objective was life, understood as the basic needs, man's concrete essence, the realisation of his potential, a plenitude of the possible... life as a political object was in a sense taken at face value and turned back against the system that was bent on controlling it. (1978:144–5)

Clashing political movements nonetheless agreed that politics is a matter of life: that the right to power is staked upon one's capacity to maximise and manifest life, upon one's capacity to foster life and upon one's own vitality, manifesting life (transcending the limits of the liveable). Socialist discourses contested the vitality of the bourgeoisie and claimed that it is labour and the effort of the working classes that are the vitality of society. Marxists claimed that the bourgeoisie, once vital, was becoming redundant with the completion of the process of proletarianisation, the creation of the class that is the true 'spirit', the life force, of history. Feminists characterised the male dominated state as degenerate and staked the rights of women to power upon the capacity of women to care for life as well as to create it, arguing for the creation of the 'mother state' (as will be illustrated in Chapter 5). The liberal bourgeoisie would continue to stake their own claims to power upon discourses of health and vitality – claiming to be best able to foster life through their scientific and economic prowess, and claiming to manifest life in their well fed, athletic, sexually repressed bodies. To manifest life, to secure life, to maximise life – these are now the stakes and strategies of political struggle. (Bio)politics is about potential not the past.

The establishment of bourgeois hegemony in Foucault's genealogy of biopolitics is, then, less about the establishment of one group's domination over others than it is about the production of a new technology of power and new formation of experience. In *The History of Sexuality: 1* and *Society Must Be Defended* Foucault suggests that this new formation of experience and the general establishment of biopolitical power is, amongst other things (such as the secularisation of the Christian pastorate), the generalisation of bourgeois power and embodiment. Foucault discusses the movement whereby technologies and discourses that were originally developed in the self-affirmation of the bourgeoisie were adopted in anti-bourgeois socialist politics (which, Foucault claims, never challenged the biopolitical basis of power (2003b:261)). Arguably this history also pertains to nationalism; modern nationalism can be understood as a kind of generalised version of bourgeois self-affirmation and class-racism. Indeed, Foucault identifies the origin of the universal statist nation in the class history of the bourgeoisie (Ibid.:227–37).

The generalisation of bourgeois embodiment

In *Society Must Be Defended* Foucault discusses the establishment of the nation state through reference to the discourses of bourgeois historians. In the eighteenth century the aristocratic concept 'nation' did not pertain to a group of people joined by a particular territory but rather to potentially mobile, familial-clan type entities. Nations as they had been defined in aristocratic discourse were always plural. The ruling aristocratic nation was defined in relation to other nations living on the same land, vanquished or – as in the writings of Boulainvilliers which decried the loss of aristocratic power – victorious nations. In the eighteenth century, the bourgeoisie claimed (as did the aristocrats) to be *one* nation amongst a plurality. At the time that they were busy producing for themselves a class body (a sexuality, a family body, a medicalised body) the bourgeoisie conceived of themselves as *a* nation amongst others. Once their hegemony was established in the nineteenth century, however, the relationship between the nation, State and conquest was overhauled (Foucault, 2003b:215–37).

In the historical discourse of the eighteenth century, the State is not (yet) the nation state, it is the sovereign government of a territory upon which might reside many different nations and peoples. The aristocratic nation based its claims to the State (to the right to govern the territory and other nations) upon past conquest. Aristocrats ruled as victors; as such, their very claims to power assumed that there must be other vanquished nations, and that the State must participate in their ongoing domination. The aristocratic technology of power implies plurality. The bourgeoisie, however, staked *their* right to govern upon their *present capacities* – not upon their past conquests or alliances. The bourgeoisie claim the right to govern because they are the nation that is most capable of fulfilling the capacity and destiny of the State. Staking their claims to nationhood upon present capacities means that the nation comes to be defined in relation to the State (to the present power) rather than to external nations (to past conquest); 'Not domination, but State control' (Foucault, 2003b:223). The organisation of national power upon present capacity has a unifying potential – power as *present capacity* does not have to be defined through a relationship of enmity. The biopolitical orientation of the grid of historical intelligibility and values upon the present and potentiality accompanied the establishment of national unity – the idea that there is *one* nation on a sovereign territory which is represented and unified in the State.

Once bourgeois hegemony was established in early nineteenth century France, according to Foucault, the bourgeois historians claimed that all the dualities and pluralities of power and people had been overcome; 'henceforth', it was enthusiastically declared by bourgeois historians, 'there will be *one* nation, *one* people and the State will be at the centre maintaining a direct relationship with – and thus between – everyone'. Foucault quotes Augustin Thierry who claimed, in 1820, that his present moment was witness to an immense evolution 'which causes all violent or illegitimate inequalities – master and slave, victor and vanquished, lord and serf – to vanish one by one from the land in which we live. In their place, it finally reveals one people, one law that applies to all and one free and sovereign nation' (Thierry cited in Foucault 2003b:236). Foucault identified the establishment of discourses of national unity with the biopolitical formations of experience, the orientation of time upon the present potential and thus life. He wrote:

> In the history and the historico-political field of the eighteenth century, the present was, basically, always the negative moment. It was always the trough of the wave, always a moment of apparent calm and forgetfulness... And now we have a very different grid of historical intelligibility. Once history is polarized around the nation/State, virtuality/actuality, functional totality of the nation/real universality of the State... the present becomes the fullest moment, the moment of the greatest intensity, the solemn moment when the universal makes its entry into the real. It is at this moment that the universal

comes into contact with the real in the present (a present that has just passed and will pass), in the imminence of the present, and it is this that gives the present both its value and its intensity, and that establishes it as a principle of intelligibility. (Foucault, 2003b:227)

In the works of nineteenth-century bourgeois historians we can trace a movement from bourgeois class self-affirmation to the totalising self-affirmation of the modern nation. Of course national unity was not in fact established, conflicts and inequalities were not, in fact, overcome in modern nation states. However, we can, arguably, see in Thierry's assertions the emergence of one of the most significant political discourses of the past century, the discourse of nationalism. Arguably, modern nationalism can be seen, historically, as the generalisation of bourgeois self-affirmation and class racism.

In *The History of Sexuality: 1* Foucault discusses the generalisation of bourgeois class embodiment to the whole of society with respect to sexuality (1978:103–31). As has been said, Foucault insists that sexuality was first developed by the bourgeoisie *for the bourgeoisie,* as an aspect of its self-affirmation and class distinction. Foucault does not argue that sexuality and the embodiment that it engendered have always been restricted to the bourgeoisie, only that it was the bourgeoisie that first developed sexuality for itself. After a series of political struggles and economic emergencies sexuality was extended to the whole of society. With this extension came an *internal* differentiation of the population body, manifest in distinctions between *types* of sexuality. Once sexuality was extended to the proletariat it became internally differentiated. Foucault states that once the proletariat and bourgeoisie both have a sexuality, sexuality distinguishes between groups on the basis of degrees of repression. 'Henceforth social differentiation would be affirmed, not by the "sexual" quality of the body, but by the intensity of its repression' (Foucault, 1978:129). Arguably this extension of sexuality can be seen in terms of a *generalisation* of bourgeois power, embodiment and nationhood. Henceforth the sexualised population would include the whole of society. Instead of distinguishing itself from other classes it would be *internally* distinguishing – a whole biological national body, with *biologically* distinguished internal fragments. It was, then, no longer a question of sexuality constituting the embodiment of one particular nation, the bourgeoisie. Rather sexuality was constituting the embodiment of a *single* population, *the* nation, an embodiment that was unified with and by the State, and that was internally differentiated through, essentially racist, biological type divisions and relations.

Foucault did not, himself, extend his explorations of bourgeois power into an explicit analysis of the experiential economy of nationalism. However I would suggest, by way of an aside, that such an analysis is one of the great potentials of Foucault's theories of biopolitics for the sociology of contemporary politics. At the least, such an experiential economy of nationalism would recognise the close relationship between sexuality and productions of trans-organic embodiment and nationalism. It would also recognise that such trans-organic embodiment is established through discourses of vitality, potentiality and creativity, and that nationalist racism has a necessary relationship to *life,* rather than, as has often been assumed, the somatic, determinist or essentialist. In an experiential economy of nationalism, that drew on Foucault's insights, life, potentiality and creativity would be understood as more essential to nationalist discourse and its racisms than would notions of fixity, essence or prexistence. The transformation of limits, rather than conservative experience, would be posited as necessary to the affective strength of nationalism in a biopolitical economy.

Biopolitical embodiment

Returning to our present project, what is most important from the point of view of our interpretation of Foucault's theory of biopolitics as experience is the link between modern political embodiments, the augmentation of peoples' embodied capacities and the political rhetoric and strategisation of a politics and ethics of life. Modern political bodies are, following Foucault, neither ideological illusions nor juridical metaphors. They are real embodiments, real positivities, real experience. They are vital forces, transforming, breaking limits, processes of life – not imaginary identities or organic entities that exert or manifest static order (engendering *conservative* experience). The capacities of peoples' bodies are *invested* in these embodiments; the vitality of bodies are subject to determination but they are also expanded, augmented, intensified. The politics of biopolitical embodiments is a politics of *positivity.*

In recounting the history of bourgeois power, Foucault establishes a link between the modern nation state, sexualised embodiment, the orientation of history upon the present passing moment, and a political affirmation of life as object and objective of politics. Biopolitical embodiment – the class, the nation – is a constellation and production of capacity, potentiality and life. It is produced through technologies such as sexuality. The force of sexuality as a discourse and apparatus is vested in its capacity to make matter matter, to make the (previously) private and mundane issues of the body into issues of immense public significance. Sexuality makes bodies responsible, making them subject to political influence and the subject of, the agent of, political influence and public significance. The generalisation of sexuality and the widespread production of biopolitical embodiment constitutes a generalisation of historical and political importance and power. It makes bodies matter more. Biopolitical embodiments (classes or nations) augment the forces of bodies of biopolitical subjects, enfolding those subjects' bodies into and around the future. In the context of biopolitical experience, potential historical significance is no longer the preserve of the very few made manifest in public spectacle. It is the common property of every body and it is made manifest in the (previously) private domains of corporeality and care. Struggles over specifically biopolitical power have been conducted within a common framework of experience and evaluation; they have been struggles to manifest, protect, secure and maximise present life, whether the protagonist in the struggle was the bourgeoisie, the proletariat, socialists, feminists, nationalists, doctors, whoever. Biopolitics is a politics of the passing moment, the transformations, the transcendence of limits, that make the present pass.

Before concluding this chapter we will briefly consider the issue that is sometimes considered a paradox of biopolitics. Biopolitics is a politics of life and yet, seemingly paradoxically, it is strongly associated with politics of death, with concentration camps, genocide, forced sterilisation. Foucault's analysis of biopolitics, grounded in a detailed grasp of the biological conception of life (to which death is internal), demonstrates that this is not, in fact, at all paradoxical. Contra Giorgio Agamben (1998), the intimacy of death to the politics of life does *not* mean that biopolitics turns into a politics of negation, deduction or death. Biopolitics is always a politics of life and of positivity according to Foucault. What it *does* mean is that racism is produced and required as an essential technology in the biopolitical 'tool kit'.

Racism, death drives and the biological-type relationship

We have seen that biopolitical power augments the forces of people's bodies, vitalising bodies, and that biopolitical discourse deploys a rationality, or set of reasons, that describe the purpose of power as the maximisation or protection of life. To grasp the force of Foucault's analysis of modern biopolitical experience, and the experiential economics of biopolitical power, it is essential to realise that biopolitics is *always* a politics of positivity – producing experience, life, augmenting and intensifying capacities – even in its darkest, most 'negating' moments.

Much of the interest in theories of biopolitics in present scholarship, particularly in the context of political theory, is concerned with themes such as securitisation, concentration camps and eugenics – themes that have been described by Marianna Valverde as 'negative biopower' (2007:176). This focus in the contemporary literature is due, in significant part, to the influence of Agamben, who has tied the concept of biopolitics to concentration camps, states of political exception and the production of the condition of 'bare life' (Agamben, 1998; 2005). We will discuss Agamben's interpretation of Foucault in the following chapter. The focus also derives, however, from Foucault's own potent remarks. For example, in *The History of Sexuality: 1* Foucault stated, of biopolitical power, that:

> Wars are no longer waged in the name of a sovereign who must be defended; they are waged on behalf of the existence of everyone; entire populations are mobilised for the purpose of wholesale slaughter in the name of life necessity: massacres have become vital. It is as managers of life and survival, of bodies and the race, that so many regimes have been able to wage so many wars, causing so many men to be killed. (1978, 137)

Although Foucault certainly did not see Nazis and concentration camps as the typical form or nomos of biopolitics (in contrast to some contemporary philosophers of biopolitics (Agamben, 1998; Esposito, 2008)), he did tie the theme of biopolitics to the explanation of Nazism, 'negative' eugenics and genocide, linking the racist politics of the Nazis to Darwin's theories, biological knowledge and experience (Foucault, 2000k:358; 2003b:259).

Foucault describes and analyses biopolitics in terms of a productivity of life and experience – in terms of positive power. Nonetheless, the perpetration of death and the 'murderous functions of the state' are central to his concerns (Foucault, 2003b:256). Indeed, the account of biopolitics in terms of positive power is a *critical* account – it is an exercise in positive *critique*. A key target of that critique is the murderous functions of the biopolitical state, represented historically in eugenics and colonialism.

Valverde remarks that there are two 'faces of biopower': a positive, life producing face, and a negative, murderous face (2007:176). However, this is misleading, making it seem as though biopolitics is no longer a play of positive power when it orchestrates the murderous functions of the state. The point of Foucault's analysis of the place of death in biopolitics is that biopolitical technologies are capable of making the perpetration of death – physical or political – be experienced as a maximisation of life, an augmentation of capacity and experience. The originality of Foucault's analysis of death in biopolitical power is to demonstrate that there is, in fact, no contradiction between maximising life and perpetrating death. There is not in fact a paradox, and there are not 'two faces' of biopolitics, one negative and one positive. Rather there is one positive power of life, biopolitical power, which is sometimes exercised and manifested through perpetrations of death.

Foucault's understanding of the capacity of death to constitute the experience and perpetuation of life is grounded in his detailed grasp of the historically specific conception of life and biology that is at play in modern evolutionary thought. Biological life, as we have seen, is not the vitality of living organisms; it is trans-organic and it is perpetuated at the limits of organisms, in sex and in death. Knowledge of death can constitute experience as life.

In the context of the biopolitical state, what makes it possible to incorporate the perpetration of death into the maximisation of life is, according to Foucault, the introduction and generalisation of a new form of racism. What makes it possible for a technology of power that takes life as both its object and its objective to exercise the power to kill, as well as the indirect forms of murder that are integral to the functioning of the State, is racism (Foucault, 2003b:254). The emergence of biopolitics inscribes racism 'in the mechanisms of the State' (Ibid.).

A specifically modern form of racism emerges with biopolitics, a racism which, like biopolitics, is organised and formulated through biological knowledge. Foucault is not claiming that racism was invented at this moment in history. But he does suggest that a new type of racism developed at this point and claims that it was at this moment that racism became *technologically* essential to the functioning of state power. This racism was formulated in the context of biological knowledge, a knowledge that is less concerned with superficial classification than it is with the underlying logics of evolutionary life. Building upon the differences between biological knowledge and natural history set out in the previous chapter, we can infer that specifically biological, biopolitical racism is a racism that is concerned with historicity, connections and dynamism, not with a fixed order of being. Biopolitical racism is not racism which is concerned to establish the facticity or validity of a fixed order in the world, ordained either by nature or by God. This is not to deny that such racisms exist and are significant. They might be associated with natural history and disciplinary power.

But they are not the forms of racism that operate within biological knowledge and biopolitical technologies. They are not the types of racism that gave shape and legitimacy to eugenics. Biological, biopolitical, racism is concerned with the ways in which bodies are *dynamic*, evolving and empowered, and it is concerned with the ways different bodies and different races are *connected*. It is an *inclusive* and *dynamic*, historicising racism.

What this racism achieves is to *fragment* the biological field that power controls (Foucault, 2003b:255). The population is divided, through racism, into a mixture of races or 'subspecies', divisions which introduce a break between 'what must live and what must die', in order for the life of the species, the population as a whole, to persist or grow (Ibid.:254). The different fragments are not defined as different populations but as variations within *one* population, fragments that are defined by specifically biological (which is to say, trans-organic, not phenotypical) differences – differences in degrees of sexual repression, in health, in degeneracy. Crucially, this fragmentation of life makes it possible to establish a *positive* relation between death and life. The death or manipulation of one fragment of the population might enhance or multiply life itself, the trans-organic life of the population or species.

The positive relation between life and death is familiar from the context of war and the relationship of enmity. In war it is normal to kill one's enemy so that one can live oneself. However, biological knowledge and biological racism do, according to Foucault, establish a different relationship. This positive relation is not about protecting a self from an Other. It is about purifying and improving 'me' by working on, eliminating parts of, the life that I most fundamentally am. Foucault describes the relationship thus:

> The more inferior species die out, the more abnormal individuals are eliminated, the fewer degenerates there will be in the species as a whole, and the more I – as species rather than as individual – can live, the stronger I will be, the more vigorous I will be. I will be able to proliferate. (2003b:255)

The relationship that is established in this specifically biological formation of racism is perhaps a more dangerous racism than 'ethnic racisms', which pertain to relations of enmity and otherness. For this biological-type relationship is placing, precisely, a *positive*, vitalising value upon death. This racism does not simply *permit* the perpetuation of death (in the name of our own security and protection, or as a result of the inhumanity and unworthiness of the racialised other). It *encourages* the pursuit of death as a means of manifesting life. Foucault continues:

> The fact that the other dies [in the biological-type racist relation-ship] does not mean simply that I live in the sense that his death guarantees my safety; the death of the other, the death of the bad race, of the inferior race (or the degenerate, or the abnormal) is some-thing that will make life in general healthier: healthier and purer. (Ibid.)

Foucault is explicit that this is not the political relationship as defined by Carl Schmitt, the 'military, warlike, or political relationship', but a specifically 'biological-type relationship', a relationship that pertains to internal differences, that is produced in the context of a normalis-ing, regularising power which produces differentiations as a part of itself.

Although this specifically biological racism did, according to Foucault, develop *first* in the context of colonisation and colonising genocide, it is by no means restricted to ethnic or phenotypical dif-ferences (Foucault, 2003b:257). Biological racism is essential to tech-nologies of biopolitical states and comes into play whenever they are faced with the need to deny life, either directly or indirectly – 'expos-ing someone to death, increasing the risk of death for some peo-ple, or, quite simply, political death, expulsion, rejection and so on' (Ibid.:256). Racism is deployed with respect to ethnic groups, social classes, the insane, to criminals, to the poor. Foucault argues that evo-lutionary biology, or 'a bundle of notions' that were loosely drawn from it, including 'the hierarchy of species that grow from a common evolutionary tree, the struggle for existence among species, the selec-tion that eliminates the less fit', rapidly became, in the nineteenth century:

> not simply a way of transcribing a political discourse into biologi-cal terms, and not simply a way of dressing up political discourse in scientific clothing, but a real way of thinking about the relations between colonization, the necessity for wars, criminality, the phe-nomena of madness and mental illness, the history of societies with their different classes, and so on. (Ibid.)

In Chapter 5 we will see that this analysis can even extend to the femi-nist confrontation with patriarchal domination. 'Whenever', Foucault continues, 'there was a confrontation, a killing or the risk of death, the nineteenth century was quite literally obliged to think about them in the form of evolutionism' (Ibid.).

Alongside the generation and generalisation of biopolitical embodiment, with all its positivity, all its investing and augmenting of bodies' capacities, making matter matter, was the generation and generalisation of biological racism as an essential component in the technological arsenal of biopolitical power. That racism not only *allows* biopolitical power to perpetrate death, removing a block upon the murderous functions of the state. It can also place a positive, life-maximising, value upon the perpetration of such death (political or physical), enabling power to manifest the maximisation of life through the pursuit of death. The politics of life and creativity can generate a *drive* to death and exclusion as an alternative (and often much easier) means of pursuit of the maximisation of life.

Biological politics, even at its most deadly, most essentialising and most life-denying remains, following Foucault's analysis, a politics of positive power, manifesting life, creativity, vitality, the transcendence of limits, the passing away of time.

Conclusion

We have seen that biopolitical trans-organic embodiment corresponds to a radical increase in bodily forces – as each somatic body becomes a constellation of the capacities of a potentially infinite series of others. In the biopolitical context – the context of modern sexuality – embodiment transcends bodies. This augmentation of forces expands and engorges the present within history. With all this connection and multiplication of embodiment there is quite simply an *awful lot more going on* – the present becomes more dense, more intense. The distant past and the realms of eternity become relatively pale next to a passing moment and imminent future, which have become immeasurably more full, intensive and creative. We can describe this transformation as a 'positive event', in the sense that it increases what is present in the world, multiplying forces. Those forces are not, however, all positive, expanding, joyful forces. The increase in power that is effected through the production of trans-organic embodiment is not a moment of empowerment so much as a transformation of the fields of affective and epistemic intensity, a transformation in the dimensionality of experience – irrevocably altering the *stakes* of politics and ethics, radically reorienting the spatio-temporal location of truth, significance and value.

Sexuality and the production of trans-organic embodiment multiply and extend the capacities of organic bodies. This does not mean, however, that those organic bodies, those people, are more empowered. We have seen that bodies become more significant within the ontological framework of sexual reproduction and trans-organic embodiment. People's bodies matter a great deal more because they *are* the matter of other people's bodies. The forces of bodies are multiplied as they come to be understood as connected through processes of creation, protection and disease. Individual bodies are, in this sense, more powerful – they have more (worldly) capacities. It would, however, be misleading to describe biopolitical bodies as more 'empowered'. Somatic bodies apprehended within biopolitical technologies and discourse (such as modern sexuality) have an intensified capacity by virtue of their biopoliticisation – they are more *significant*. However, an intensified *significance* does not correspond to greater empowerment. That bodies are more significant from the point of view of governance and history also means that bodies are more subjectable: that there is greater cause and legitimacy for the interference with other people's bodies. We can more accurately say, then, that the production of trans-organic bodies constitutes an *augmentation of embodiment*, incorporating somatic bodies into more numerous fields of intensity and affect, making bodies *more embodied,* insofar as embodiment refers to constellations of capacities to affect and be affected rather than to physical being. Biopolitical organisation of bodies makes any given somatic body a constellation of more capacities. Those capacities might, however, be depleting and dissipating just as readily as expanding and intensifying.

The constitution of trans-organic embodiment is a radical affective event. The generation of trans-organic embodiment transforms the ground of affective investment, epistemology, the formations of experience and thus the possibilities in the formations of values and meaning. The production of trans-organic embodiment does not only constitute an increase in the expansion of forces (and thus of joy and joyful attachment); it also constitutes an increase in the depletion of forces (and thus of sadness and negative investments – fear, depression, resentment). The biological, biopolitical refiguration of experience does not necessarily make people more happy – increasing the expansive forces of bodies. Rather, it produces a new terrain of emotion, a new plane of affective investment. When the 'social body' changes from a mere metaphor to an actual (corpo)reality, society (as other individuations of trans-organic embodiment) can become a focus for affective investment and thus for political, ethical and epistemic activity and intelligibility. Whether it appears as Society, the proletariat, the race or the nation, the trans-organic body of the population can become something worth living (and killing) for.

The generation of trans-organic embodiment, radically multiplying the causal connections between somatic bodies, transforms the historical grid of intelligibility, value and meaning – orientating duration upon the present and passing moment. It produces collective embodiment as an intensely affective reality in which individual and family bodies are themselves invested. It generates a whole new world of stakes, ambitions and responsibilities for political power and struggle. The generation, or recognition, of trans-organic embodiment 'unblocks the art of government' (Foucault, 2007:104–5); it de-privatises corporeal and family life, giving life itself over to the realm of politics, and demanding of politics that it base its claims and values upon the protection and maximisation of this life.

With the production of trans-organic embodiment, the present and immediate future becomes full of potentiality and significance. Relative immortality is present within the living world as manifest in population life. The mere fact of living becomes the manifestation of *life* – a process that is self-transcendent, (quasi)-transcendental, a becoming more than itself. In the following chapter it will be argued that Foucault, with Hannah Arendt, characterises the modern population life as something like the immanentisation of the eternal soul of Christian theology. Foucault and Arendt also concur in placing an emphasis upon the experience of processuality in the formation of modern political values. The key argument to take forward from this chapter into the next is that this immanentisation of eternal life or 'the soul', and this processualisation of present experience, are the result of a *collectivisation of embodied potentiality*. Vitalist biopolitical values – values that are orientated upon maximising, securing, purifying, regulating and, above all, *manifesting* life – pertain to life as a process of self-transcendence and escape from finite singularity; a *this worldly*, immanent, movement of forces beyond present limits. Even individualistic pursuits of health, creativity and vitality can be seen as participating in this attempt to escape from finite singularity; manifesting, embodying, a vitality that always escapes from and transcends the body in which it is fixed for a moment.

Modern sexuality organises bodies as participants in flows of life that are trans-organic, reaching beyond and breaking the limits of given organic bodies. Whilst there is a clearly subordinating movement in this, there is also a movement of augmentation and production, increasing the capacities and significance of people's bodies. The success, the hold, of sexuality as power can be understood in terms of this productivity – a production of experience. Biopolitical embodiments, modern classes, nations (and, as we will see in Chapter 5, politicised genders) are animated by this experiential economy, the augmentation and investment of bodies' capacities. Crucially, for Foucault, that positivity is at play even in the production and performance of life-negating politics. Racism enables the perpetration of political and physical death to be experienced as the maximisation of life.

3

Christianity, Process and Positive Critique: Rethinking the Resonance between Foucault and Arendt, against Agamben

Giorgio Agamben's writings have been enormously influential in the thinking of biopolitics over the past decade, especially in the context of political theory. Clearly Agamben's description of biopolitics finds resonance in the horror of a moment steeped in the souped-up securitisation and displacement of 'crisis', 'emergency' and 'post 9/11' or 'post Cold War' geopolitics. For all his pertinence, however, Agamben's analyses of political process seems strangely removed from political *experience*; from economies of bodies, desires and connections; from the subjective eventful practical stuff of political life. As such, Agamben's interpretation of biopolitics points away from some of the most valuable and engaged aspects of Foucault's thinking on these themes.

This chapter will begin by problematising Agamben's account of Foucault's ideas on biopolitics, drawing out different interpretations that result from focusing on biopolitics as *experience*. It attempts to go beyond a *negative* critique of Agamben's work and to develop a more 'positive' critical response to Agamben. To do this, the chapter takes up a constellation in thought that Agamben identifies, and from which his arguments on biopolitics set out. That constellation is the commonality between Foucault's work on biological politics and Hannah Arendt's reflections on the human condition in modernity. I will push that constellation in different directions than does Agamben and work towards an alternative reading of the resonance and tensions between Foucault and Arendt. The intention, then, is to 'maximise' the topic that Agamben addresses – to push beyond the limits he has drawn, producing something new – rather than rubbishing his position. We will start from the point that Agamben is *correct* to situate Foucault and Arendt together.

Foucault and Arendt can be read together insofar as they are both *positive critics* of biological politics. They both address the positivity of biological politics and knowledge. Whilst holding biological knowledge in part responsible for some of the most terrible atrocities of the past century they, nonetheless, attempt to understand its genuine *appeal*. Both emphasise the relationship between biological thought and *processuality* (although that has different implications for each of them). Both also emphasise the place of Christianity in the genealogy of biological-political ideas and values. The biological life that is the centre of modern politics can, according to both Foucault and Arendt, be seen as something like an immanentised, secularised, massified version of the eternal soul that was at the centre of 'pastoral power' or Christian teaching. Nonetheless, there are, as the chapter will show, important differences between Foucault and Arendt. These pertain to the relationship between individualisation and normalisation, or individual freedom and political power. Whereas individuation and normation are effectively opposite in Arendt, they are part of the same process as Foucault sees it. This difference explains why for Arendt the political can only take place in the public, whilst for Foucault work on the limits of the self, ethics, work in private, can be of public, political significance.

From the perspective of the project of this book it is important to work upon Agamben's reading of Foucault and Arendt, to push it into another direction, because the picture of 'what biopolitics means to Foucault and Arendt' that Agamben puts forward radically eschews the character of the experience that is produced through biological and biopolitical knowledge. Agamben's analysis would entail either that biopolitics is rationalistic and destructive of experience, reducing Foucault's analysis to the Weberian style of critique, or that biopolitical experience is conservative, an experience of fixity and stability not the transcendence of limits. Either view obscures the experiential dynamics of biological life as Foucault understands it and confuses the economy of biopolitical experience that is being developed here.

Although there is much of value in Agamben's independent arguments about biopolitics, his presentation of Foucault's and Arendt's theses on the (bio)politics of modernity is hugely problematic. The transhistorical nature of his analysis and focus upon sovereignty, law and philosophy is completely out of keeping with both Foucault and

Arendt – and thus the points that they make are massively distorted when transcribed into Agamben's analysis. The thrust of at least some of Foucault's arguments is all but inverted in their representation in *Homo Sacer,* wherein Foucault's *bio*-politics *for life* becomes Agamben's *thanato*-politics *for unity and order.* Whilst these problems are less acute with respect to Agamben's reading of Arendt they are, nonetheless, present.

Whilst I disagree with Agamben's presentation of Foucault's and Arendt's arguments, I do not disagree that there are significant resonances between these two thinkers. The resonance is *not,* however, some magical point of triangulation which grants us access to the otherwise secret trans-historical truth of Western political reason. I propose, instead, that the resonances between Foucault's and Arendt's arguments be understood genealogically, in terms of their relation to the mainstream European/American left with which they were contemporary. The critique of biological thinking was central to post-war politics and social science. Being interested in the role of biological thinking in modern politics did not, then, mark Foucault and Arendt out as unusual. What differentiates them is the attention that they pay to the *positivity* of biological thinking – the fact that they were committed to understanding biological thinking from 'the *inside*'; to understanding its genuine appeal. Rather than equate biological thinking with its negative effects for the oppressed – with its determinism and conservatism – Foucault and Arendt describe biological thinking in its *own* terms, in terms of what it does itself say it values: *life, health, evolution.* From this perspective biological thinking is about intensity, expansion, perpetual transformation and process – health, excellence and vitality. The life to which it is addressed can be understood as something like an immanentised version of the eternal life for which Christian ethics and pastoral power had previously taken care. Biological thinking might have – indeed *does* have – powerful conservative and thanato-political *effects* but it couldn't generate those effects, it wouldn't have the force it does, were it not many other things besides.

The embodied, emotive and aesthetic dimensions of biological thinking, and thus the deployment of ethics in biopolitical governmentality, are located in the positivity of the biological. This is obscured by the rendition of Foucault's and Arendt's arguments in Agamben's *Homo Sacer,* which refocuses the critique of biological thinking back upon ideas – philosophy and law – and ignores the embodied, emotive, aesthetic and ethical sides of biological thinking: the positivity that Foucault and Arendt both identify as they move *beyond* the idealistic accounts of the political, as expounded by ideology critique, *towards* a genealogical account that begins with positive forces, struggle and embodied strivings for power.

Agamben is closer to Arendt in his assessment of modern politics than he is to Foucault, and this nexus marks a contrast and intensive tension *between* Arendt's and Foucault's ideas. Arendt and Foucault differ in their assessments of the relationship between individuation and the norm, or normalisation, within modernity. Whereas the normalising force of modern society is opposed to all individuality for Arendt, individuation is a part of modern normation for Foucault – or rather, the force and authority of community and norms is enfolded within individuation according to Foucault. Individuality thus occupies a very different place in the assessment of modern (bio)politics, as well as in its potential opposition, according to Arendt's and Foucault's assessments.

Agamben on biopolitics

Agamben commences *Homo Sacer* by introducing a distinction between different conceptions of life that were present in Greek philosophy. The Greeks had no single term for 'life'; rather, they had two distinct terms, one pertaining to the simple fact of living, bare life and the other pertaining to something like 'the way of life'. The term *zoē* 'expressed the simple fact of living common to all living beings (animals, men, or gods)' whilst *bios* 'indicated the form or way of living proper to an individual or a group' (Agamben, 1998:1). In Classical thought matters concerning *zoē*, and thus matters of life's survival, were excluded from the domain of politics – the *polis*. The concerns of *zoē* were to be dealt with in the privacy and despotism of the home – the *oikos*. Man was for Aristotle a living animal with the *additional*, strictly separate, capacity for political existence.

Agamben makes much of Foucault's reference to Aristotle in the conclusion to the *History of Sexuality: 1*, wherein Foucault states that modern man is, to the contrary of what he was for Aristotle, 'an animal whose politics calls his existence as a living being into question' (Foucault, 1978:143; Agamben, 1998:3). Biopolitics is what happens when political man becomes concerned (and obsessed) with his existence as a natural, living being, and 'the species and the individual as simple living body' becomes what is at stake in political strategies (Agamben, 1998:3). Foucault's biopolitics, and man's biological modernity, is – according to Agamben's reading – the entry of *zoē* into the concerns of the *polis* and the subjection of *zoē* to the political *technē* of sovereign power.

This is a de-historicising move on Agamben's part in a dual sense. First, Foucault's comments are removed from *their* context, which is to say from a genealogy of sexuality that is concerned with the constructed and historical status of any pre-political notion of physicality (foreclosing any *trans-historical* distinctions such as *zoē/bios*, bare-life/human-life, or nature/culture). Second, the historical specificity of notions that are central to biological thinking, such as species, is obliterated, whilst all thought of living physicality is subsumed under a 'mere' physicality, 'bare life', that one could write of equally well in the ancient *polis* as one could today. Agamben, then, transforms Foucault's thickly historical, genealogical concept of 'biological life' into an abstracted concept referring to a presumed transhistorical category – *zoē*.

This de-historicisation is in a sense deliberate, in that Agamben argues that Foucault was *mistaken* in tying biopolitics exclusively to modernity. Foucault, he seems to suggest, effectively bought into an Aristotelian fiction and believed that until modernity the public domain of politics had restricted itself to questions concerning the *bios* and the good life, relegating matters of mere survival – *zoē* – to a realm beyond politics. Agamben *himself* sees through this fiction and will claim that biopolitics is something like the secret truth of all Western politics, law and political-philosophising. Contrary to Foucault's (supposed) thesis, *zoē* has been included in the *polis* of western politics all along, since its ancient inception. *Zoē* is included as the necessary exclusion, the state of exception, that (inversely) effects the inside of the *polis* and sovereign power. Bare life – *zoē* – is the *necessary* object of sovereign power. 'The fundamental activity of sovereign power is [and was since antiquity] the production of bare life as originary political element and as threshold of articulation between nature and culture, *zoē* and *bios*' (Agamben, 1998:181). Without bare life (the state of exception of political right) there is no sovereign power. Thus political *technē* has always had to engage in the delimitation and subjection of *zoē*.

This is not to say that there has not in fact been a radical transformation of biopolitics in modernity according to Agamben. What *does* distinguish modernity, for Agamben, is not the entry of *zoē* into the realm of the *polis* as such, nor the introduction of *zoē* as an object of power (these are ancient phenomena), but the becoming indistinguishable of *bios* from *zoē*. The 'realm of bare life – which is originally situated at the margins of the political order – gradually begins to coincide with the political realm, and exclusion and inclusion, outside and inside, *bios* and *zoē*, right and fact, enter into a zone of irreducible indistinction' (Agamben, 1998:9).

According to Agamben's analysis, there is an immensely reductive force in this modern moment. The political life, proper to the human of the city (culture), is reduced to the bare, 'biological' life of animality (nature) – which, most troublingly, is also the life that is exempted from the protections of the law and can legitimately be killed. Modern biopolitics *is* specific then, and its specificity is about the reduction of political, speaking, cultured man to the bare 'biological' life of animality *throughout* the *polis*.

It is no wonder, then, that the great totalitarian states of the twentieth century appear to Agamben as the 'exemplary places' of modern biopolitics (Agamben, 1998:119). The concentration camp is, he proposes, the ' "nomos" of the modern world. Today it is not the city but the camp that is the fundamental biopolitical paradigm of the West' (Ibid.:181). For Agamben any treatment of biological life is treatment *as* bare life. There is no positivity in the biological. And all biopolitics, from the elimination of Jews, through the purification of the biological body in the elimination of the mentally ill, to the elimination of the poor classes through economic development... all are reductive, eliminative, thanatopolitical – all transform the object of power (Jews, the ill, the Third World) into bare life and seek to eliminate it (1998:179–80).

So Agamben equates 'biological life' with a kind of transhistoric, extra-political fact of living, death and survival. It is a 'bare' life that is always subjected (even when it is also subjecting). It stands in contrast to culture, to logos, to 'the way of life' and political existence proper. In this move he at once makes of biological life an ahistoric (or at least immensely transhistoric) phenomenon and (re)installs the assumption that biological life is necessarily something reductive, objectified, 'bare'. Bringing biological life into political play will then inevitably and unerringly constitute a movement of *reductive* force, if we are to follow Agamben. Biological life is an *ahistorical* category and the entry of biological thinking into the political realm is necessarily a *reductive* and objectifying process (destroying history, logos, culture). In both these respects Agamben is at odds with Foucault's understanding of biological life and thus of biopolitical configurations of power relations, embodiments and ethics.

Problems with Agamben's account of Foucault

There is doubtless much insight in Agamben's arguments about bare life and states of exception, arguments that have undergone considerable expansion in the decade intervening since *Homo Sacer*'s publication, and which shed greater insight upon the politics of the twenty-first century than the nineteenth (see Agamben, 2004; 2005). But whatever the independent merits of his thought, Agamben seriously misconstrues Foucault's ideas about biological life, giving not so much a critique as a misrepresentation of what Foucault himself has to say, occluding both the *historical specificity* and the *positivity* of the 'bio' of biopolitics.

The biological life that Foucault is talking about did not exist in the eighteenth century, let alone in the third before Christ (Foucault, 1970:127–8)! And the notion that a categorical distinction between such things as *bios* and *zoē* could hold across millennia is wholly out of keeping with Foucault's conception both of history and the thickly, intricately constructed nature of any such power/knowledge entity. The biological life that enters politics in the nineteenth century, according to Foucault, could not have been in play even a hundred years before it was in fact so, let alone in the trans-historic time of abstraction in which Agamben theorises.[1]

More specifically, biological life as an entity is contingent upon historically situated capabilities, such as the development of statistical analysis through the disciplinary eighteenth century, which facilitate the comprehension (and apprehension) of such vital phenomena as occur at the level of population (Hacking, 1982). Statistics enable the specific phenomena of population life to be recorded and thus reveal that vital phenomena are not contained in the scale of family (Foucault, 2007:104–5). It is through such technologies that man gradually learnt what is meant 'to be a living *species,* to have a body, conditions of existence [and] probabilities of life' (Foucault, 1978:142, emphasis added). Without them biological life does not in fact *exist.* Biological life is *not* any old bare, animal, physical or natural life (Foucault, 1970). It is the specific type of life to which species and populations are party, which, if not exactly discovered or invented by modern knowledge production techniques, was at least brought into view, 'carved out as a domain of reality', for the first time in the eighteenth and nineteenth centuries (Foucault, 2007:93).[2]

There is, of course, little ambiguity in the fact of a difference between Agamben and Foucault when it comes to the question of the historical specificity of biopolitics. As we have seen, the former explicitly critiques (or 'completes') the latter on this point, claiming that Foucault failed to notice the trans-historic nature of the biopolitical (Agamben, 1998:9). My point is, however, that in 'trans-historicising' the life that enters politics, Agamben radically misrepresents *what* the life that Foucault is talking about actually is, forcing Foucault's words into arguments that have very little to do with – are in some sense *opposed* to – his own.

Agamben *wants* Foucault's treatise on biopolitics to be about the reduction of culture to nature, humanity to animality, and the generalisation of an (excepted) state of totalitarian rule. Agamben *wants* Foucault to move on from his initial claims about biopolitics in *The History of Sexuality 1* to a discussion of totalitarian rule and concentration camps. This, Agamben claims, is what we might have 'legitimately expected' (Agamben, 1998:119). That Foucault does not discuss what Agamben expected, however, is indicative that Foucault is not, and never was, talking about the same politics of *zoē* that Agamben is trying to get at.

Unfortunately Agamben does not discuss the *un*expected directions in which Foucault's work actually did unfold. In fact neoliberalism, not totalitarianism, is the topic of the series of lectures to which Foucault gave the title *The Birth of Biopolitics,* whilst subsequent volumes of the *History of Sexuality* wound up investigating ethics and autonomy, *subjectifications* not objectifications.[3]

There is not simply a difference between the two on the 'when and wherefore' of biopolitics. There is a vast difference between them on the question of *what* biopolitics is. In radical contrast to Agamben and his totalitarian states of exception, Foucault is talking about a politics of life of which if anything is exemplary it is *liberalism* and in which the life in question is, in really important respects, expansive, autonomising and *positively* charged.

It is not, as I say, my intention to develop a critique of Agamben's theory of biopolitics on its own terms here.[4] The fact that Agamben is talking about a different thing to Foucault does not, in and of itself, make what Agamben is saying wrong. But it *does* make his subsumption of Foucault's arguments under his own unfortunate and inaccurate. Whatever it is that Agamben is talking about it is emphatically *not* the same thing as is Foucault.

Crucially, the biopolitical account of human existence is, according to Foucault, *subjectifying* not objectifying. Biopolitical governmentality as Foucault describes it is addressed to a world of vital, autonomous phenomena. In fact it is liberalism, with its naturalistic ideas about social and economic behaviour and autogenetic, vital, natural processes of self-regulation that must be both respected (left alone) and protected (secured) that is the archetypical form of biopolitical governance, according to Foucault. Even in its most illiberal manifestations, even in totalitarian states, biopolitical governance retains at its centre the liberal dilemma: 'how not to govern too much' (Osborne, 1996; Foucault, 2003b:202; Senellart, 2003:383–4). This is not a way of thinking about the world that treats people as objects, or as 'bare', merely surviving, living beings. Biopolitics is addressed to a world of autonomous agency; and this is really central to understanding the character and capacities of biopolitics as Foucault describes it, not least because the autonomising, subjectifying capacities of biological thinking constitute the space of ethics, providing the aesthetic and embodied cement – and producing the authority – of modern governmentality.

Agamben, Arendt and the destructiveness of modernity

The other great thinker of biopolitics in modernity, according to Agamben, is Hannah Arendt, whose *The Human Condition* traced the processes by which the *animal laborans,* and with it all previously private matters of mere life's survival, enter and destroy the public (Arendt, 1998; Agamben, 1998:3). Life-as-survival is elevated to the highest value and the creativity and courage of genuine political action is precluded in advance (Arendt, 1998:*esp.* ch.II).

Arendt does draw upon the Classical distinction between the private and the *polis* and argues that the modern age has seen the concerns and travails of labour – of life as survival – newly admitted to the public realm (Arendt, 1998:46–8). If this admission is not exactly reductive it certainly *is* immensely destructive, obliterating any vestige of a public realm, and thus any chance of genuinely creative/political action (Arendt, 1998:40–1), whilst requiring that the common world – the world that is built through human endeavour and which enables people to find meaning and worth through their efforts at artistry and perfection – is sacrificed (1998:256). The modern age has seen human matters of life and death removed from the private realm of family economy and turned into a public concern. This 'liberation' of labour effectively destroys the public/private divide and instead of families plus the public we have one 'super-human family', a biological-type entity, 'society' (1998:29; 39–40). Matters of life and death are now taken care of on a massive society-sized scale, and this isn't so much the politicisation of life as the 'lifesisation' of politics. Beauty is forgotten whilst the 'political virtue par excellence', which is to say *courage,* is rendered incompatible with a 'public' realm in which a concern with survival, the contrary of courage, has become all-consuming (1998:36).

Agamben's focus upon sovereignty, law and philosophy is hardly compatible with Arendt's 'thick' methodologies of political anthropology and genealogy. Arendt's signature method of analysis is to pluralise categories, construct minutely precise definitions and insist upon a strictly historical interpretation of concepts. She is constantly seeking to dispel (or undermine) confusions of one state of affairs with another – be that labour with work and action (Arendt, 1998), imperialism with colonialism (Arendt, 1968), or totalitarianism with authoritarianism (Arendt, 1968; 1993:91–142). It is most unlikely, therefore, that she would have had much truck with Agamben's reductive, abstracting, 'ontologising' (Dillon, 2005), account of Western political history, or with his totalising, collapsing approach to the diversity of politics of life (so many manifestations of the *same* thanatopolitical logic, recurrences with the *same nomos* – the camp).

Nonetheless, Agamben's bleak portrait of the modern biopolitical state of (excepted) affairs does sit much more comfortably with the general timbre of Arendt's writing than it does with Foucault's. Moreover, Agamben's insistence upon guarding the distinction between life as *bios* (politics, action, origin) and life as *zoē*, (labour, staying healthy and staying alive) does constitute a genuine repetition of Arendt's position. Agamben's claim to Arendt as a predecessor is, as such, considerably more plausible that is his claim to Foucault.

Recovering the link between Arendt and Foucault

Agamben's presentation of Foucault's ideas is misleading and his claim that Foucault's writings on biopolitics are a precursor to his own is inaccurate and implausible. Foucault's positive critique of biopolitics does, in fact, contrast radically with and even contradicts Agamben's conception of biopolitics as the production of enmity and homogeneity. Agamben's presentation of some of Arendt's arguments is more convincing, and there is considerably more agreement between their assessments of the state of affairs in modernity. This will doubtless lead many who share my irritation with Agamben's interpretation of Foucault to dismiss the link between Foucault and Arendt out of hand, the drawing of that link being so much associated with Agamben at the present time.

It would be a shame to do so, however, for there *are* genuine and genuinely illuminating resonances between Foucault and Arendt on biopolitical (as other) themes. Although Arendt was addressed to the problems of totalitarianism and normalisation she also, in important respects, can be seen as a 'positive critic' of biological politics. Arendt, like Foucault, was drawn to the explanation of the *appeal* of political discourses or 'ideologies', even (or especially) in their darkest and most thanato-political moments. She emphasised processuality, historicity, survival and care in her explanations of the appeal of life as value and biological labour as politics.

Agamben was not the first to draw the link between Foucault and Arendt on biopolitics, and has not done so in the most interesting or accurate of ways: Bell (1996), Braun (2007) and Dolan (2005) all give better analyses of this relationship than does Agamben. The really illuminating resonances between Foucault and Arendt concern the positivity – the historicity, productivity and experiential appeal – of biopolitics: precisely those things that Agamben obscures or ignores in their work. The tensions between Arendt and Foucault concern the issue of individuality and its relationship to normalisation and power. For Arendt these appear as opposites whereas for Foucault they are integrated.

Process, Christianity and life in Arendt and Foucault

Arendt and Foucault both associate the entry of life into politics with 'the modern age' and specifically with the development of statistics and the extended (and intensified) embodiments that they helped to carve out – 'society' for Arendt, 'populations' for Foucault. Whatever the thanatopolitical, destructive and oppressive processes and events that they will associate with 'biopolitics', or 'life in the public', both insist, with however much regret, upon its immense *positivity* ... which is certainly not to say goodness, but might be to say 'excellence', 'vitality' or even 'popular appeal'. It is on this theme – the positivity of biological thinking in the political – that Foucault's and Arendt's arguments coincide. Together they shed light on the character and functioning of ethics in relation to modern politics. Both thinkers help us to understand the possibility of that positivity as well as pointing to some of its implications for the political rationalities of the twentieth century, for totalitarianism *and* liberalism ... and socialism and feminism for that matter. The resonance between Foucault and Arendt's theses rings around the issues of processuality, ethics and the positivity of biological, or 'emancipated organic' life. *Not* around what Katherine Braun has termed 'the zoëification of life', as Agamben (and Braun (2007)) imply.

The genealogical positioning of Foucault's and Arendt's critiques of the biological

Foucault and Arendt do not share in a magical insight into some secret trans-historic truth of western society. Nor do they share in the denunciation of modernity as a zoëification of life or in a conception of the discursive frame of biology/organic life as a reductive force in politics. What they *do* share in is a critique of biological thinking in and of the political that is alert to its *positivity*. In this, both are seeking to understand the performative appeal, the play of embodiment, aesthetics and empowerment in biopolitical discourse, making an explicit effort to get away from the approach of 'ideology critique' (Arendt, 1968:7, 470–1; Foucault, 1980:118). Rather than demonstrating that such and such political discourse is 'untrue', both Arendt and Foucault, in the tradition of genealogy, seek to demonstrate where the discourse and its values have come from and what is the nature of its appeal.

Their explorations of biological thinking in the political look like attempts to engage with a framework of thought (which in fact *everyone* was in the habit of opposing and critiquing) *from the inside*, to analyse that thinking in terms of the values it itself purports to uphold and in terms of the positives – the expansions of force and embodiment – that it really does offer to those who take it up as a view of the world.

Both Foucault and Arendt illuminate seductive capacities of biological thinking in the political. Foucault relates biopolitics to medicine and to liberalism, to care for bodies and autonomisations of processes and people. Arendt, through however gritted teeth, describes the entry of labour into the public sphere as an 'emancipation' resulting in the most excellent of achievements. Public labour – the life of society – makes for so much technical brilliance in the labour of life's survival that it has transformed the entire inhabited world in but a few hundred years. Arendt writes:

> The labouring activity, though under all circumstances connected with the life process in its most elementary, biological sense, remained stationary for thousands of years, imprisoned in the eternal recurrence of the life process to which it was tied. The admission of labour to public stature … has … liberated this process from its circular, monotonous recurrence and transformed it into a swiftly progressing development whose results have, in a few centuries totally changed the inhabited world.
>
> The moment labouring was liberated from the restrictions imposed by its banishment into the private realm … it was as though the growth element inherent in all organic life had completely overcome and overgrown the process of decay by which organic life is checked and balanced in nature's household. (Arendt, 1998:46–7)

According to Arendt organic life, or 'labour', gives expression to the experience of process as value. Limitless process, process as end in itself, was, she argues, the most significant discovery of the nineteenth century, and this was, not least, because of its seductive, affective, capacities. Imperialism, for example, was such a seductive rationality and practice because it enabled businessmen and bureaucrats to feel themselves the embodiment of limitless, impersonal forces and flows – forces that expand only for the sake of their own expansion (Arendt, 1968:215). Organic life became the ultimate value in part because it seemed to be the expression of all these experiences of process; of being beyond the present singular.

The fact of the positivity, processuality, expressive and expansive character of life in the public sphere is not in any way an argument against its culpability in the reductive, conservative, objectifying and thanato-political processes that it invites or augments (contra the claim of Ojakangas, that biopolitics can only justify, not encourage, thanato-politics (2005)). The affective appeal of expansion and processuality is at the heart even of totalitarianism, according to Arendt; it is a rationalisation of governance that aims, ultimately, at the perpetuation of the party as *movement*. The negative, destructive and reductive capacities of the biopolitical are only conceivable given the positivity that it also engenders.

In contrast to an ideology-critique approach to biologism in politics, Arendt and Foucault describe biologism in terms of its positivity and capacities, illuminating the affective dynamics that give genuinely reasonable people genuine reasons to have adopted such a set of ideas and embodiments.

Process in Foucault and Arendt

In a recent article Katherine Braun claims that a key intersection between Foucault's and Arendt's theses on biopolitics concerns processual temporality (Braun, 2007). Braun's analysis of Foucault's and Arendt's ideas and their relationship is far more convincing than Agamben's. The distance between their understanding of biopolitics is marked by her insistence, contra Agamben, that Arendt's study of totalitarianism, *The Origins of Totalitarianism, is* very much concerned with biopolitics (Braun, 2007:6). This is because biopolitics should be understood, according to Braun, primarily in terms of a politics of the impersonal processes of natural law and species life.[5] I most certainly agree with Braun that *The Origins* is as fully informed by a biopolitical consciousness as is *The Human Condition.* I also concur with Braun's claim that a crucial intersection between Arendt and Foucault's thinking is a concern with a specifically processual temporality. This said, Braun's specific interpretation of what processual temporality is about is problematic, as are the inferences that she draws from this insight.

In both *The Origins* and *The Human Condition* Arendt identifies the discovery of *process* as a key event in the constitution of modernity. '[T]he concept of process', she writes, 'became the very key term of the new age as well as the sciences, historical and natural, developed by it' (Arendt, 1998:105). Processuality has a hugely significant cultural aspect for modern politics (and philosophy and poetics...) according to Arendt. As Braun notes, Arendt claims that the immersion in processuality constitutes an answer to the questions of death, finitude and loneliness in the modern age (Braun, 2007:12). And this is an answer to the condition of loneliness that is mobilised by all manner of modern political formulations, including racist ideology, imperialism and, of course, totalitarianism. Ossification is deadly for totalitarian movements, the governmental form of which is embodied in Trotsky's slogan of 'permanent revolution' and the Nazis' programme of 'racial selection that can never stand still' (Arendt, 1968:389–91). They have to keep moving. Their appeal depends upon their ability to render individuality within the present the embodiment of impersonal, transhistorical process. For Arendt, immersion in process constitutes something like the affective or aesthetic heart of modern governance, including totalitarianism.

Braun rightly notes an intersection between Arendt's thinking on process and Foucault's thinking on biopolitics, stressing the fact that the new power that Foucault describes is addressed to impersonal, *beyond-individual*, biological processes. This biopower *does not* target individuals as living beings, *does not* operate through the exercise of direct control over the body and *does not* intervene in individual lives. 'Instead', Braun writes, 'it targets collective phenomena such as the birth *rate*, or the *average* life expectancy' (Braun, 2007:11, *original emphasis*). Biopolitics is, for Foucault, about taking control of life and the biological processes of man-as-species, as population. The processes to which biopolitics is addressed take place beyond individuals' lifetimes. The birth and death of an individual is not the limit point of biopolitical governance but precisely its site of mediation and operation, such that the 'life biopolitics targets has a supra-individual dimension, not just in a numeric but also in a temporal sense' (Ibid.).[6] Foucault's analysis of biopolitical rationality comes very close to Arendt's analysis of processual thought, especially where he highlights *evolutionary* thinking as a general paradigm, 'not simply a way of transcribing a political discourse into biological terms…but a real way of thinking about the relations between colonization, the necessity for wars, criminality, the phenomena of madness and mental illness, the history of societies with their different classes and so on' (Foucault, 2003a:257).

Biopolitics, and the intersection of Foucault's and Arendt's thinking, is primarily concerned with processual temporality according to Braun. Unfortunately, however, she understands processual temporality in a limited fashion, equating it solely with the *impersonality* of biological processes and natural laws. Braun is concerned with processuality as a way of thinking about individuality that *obliterates* individuality, processes as *transcendent* phenomena that subsume individuality and individual life, operating at the point of its termination. As biopolitics is all about this transcendent processuality it is no wonder that totalitarianism is – as it is for Agamben – the epitome and highest culmination of biopolitics for Braun. Which, as we have already seen, it certainly is *not* for Foucault.

I am in agreement with Braun's claim that an attention to 'processual temporality' constitutes a crucial intersection between Arendt's histories of the modern age and Foucault's theses on biopolitics. However, much contemporary literature on processual ontology, epistemology and ethics cast this claim in a rather different light. In the literatures that are concerned with processual ethics, process ontologies and new vitalisms, process is associated precisely with *immanent* normativity, with intension rather than extension, horizontal rather than vertical transcendence (Bammer, 1991; Irigaray, 2004; Fraser et al., 2005; Lash, 2007). In the context of this literature there is certainly no necessary relationship between processuality and the subordination of individuality to transcendent, impersonal forces. Indeed, some would have processuality (wrongly in my view) associated *solely* with immanence, individuality and radical autonomisations. As such, whilst Braun is correct that processual temporality is central to the genuinely illuminating resonances between Foucault and Arendt, this does not, in itself, preclude us from thinking of biopolitics as much (or more) in terms of the problematics of individualism and liberalism as totalitarianism and totalisation.

Foucault's studies of ethics, which followed his lecture series on biopolitics, were addressed to the processes of subject formation that constitute (or enable the self-constitution of) autonomous subjects creatively practicing processual ethics. The idea of 'subjectification' developed in these studies has done much to aid contemporary thinkers in the illumination of the centrality of *immanent* normativity to modern governmentality (see Burchell, 1996; Dean,1996; Rose, 1999). An attention to processual temporality has, then, had quite different implications for Foucault and his ideas than that which Braun's presentation of processuality would lead us to expect. And even in Arendt, whose thinking on process does largely pertain to impersonal processes, there is an attention to the intensive qualities of the experience of process and a less than black and white distinction between the processual temporality of labour/organic-life and that of the political action proper and human capacities for creativity and natality (which Arendt celebrates).

The deployment of the experiential appeal of processuality does not necessarily indicate the totalising subordination of individuality to a higher force. The participation of the present in discourse, whose time is not our own (see Bell, 1996:93), means that the processual transcendence-of-self might well be *immanent* to the work of the self, its work upon itself. The processual temporality of biopolitics might, as such, pertain to the work of *ethics*, effected in and through individuality, rather than (only) to the imposition of transcendent authority, of natural laws or above-individual processes that obliterate individuality. The centrality of processual temporality to the biopolitical does not, then, tie us to a conception of biopolitics as *necessarily* or *only* totalising.

Biological life as immanent eternal soul?

A further echo between Foucault's texts and Arendt's concerns the place of Christianity in the origins of modern biopower. Foucault locates the origin of modern biopolitical governmentality in the 'pastoral power' of the Christian Church (Foucault, 2000h:332–6; 2007: *esp.* lecture 8), whilst Arendt argues that the radical value placed upon life in modern society constitutes the continuation of the Christian belief in the sanctity of life after the secular decline of the Christian faith (Arendt, 1998:313–30). Foucault's genealogy of biopolitical power traces the generalisation of pastoral power and the immanentisation of the 'eternal soul', in an argument that resonates with Arendt's claims in the final chapter of *The Human Condition* concerning the relationship between Christianity and the rise of life to the status of ultimate values in modernity.

Foucault argues that Christianity inaugurated not only a new code of ethics but a new form of power relations: what Foucault calls 'pastoral power' (Foucault, 2000h:332–4). Pastoral power: is salvation-orientated (as opposed to political-power), its ultimate aim being to ensure individual salvation in the next world; it is oblative (as opposed to the principle of sovereignty), which is to say that it must be prepared to sacrifice *itself* and not just demand sacrifice; it is individualising (as opposed to legal power), looking after each individual during his life time, rather than the whole community; and '[f]inally this form of power cannot be exercised without knowing the inside of people's minds, without exploring their souls, without making them reveal their innermost secrets. It implies a knowledge of the conscience and an ability to direct it' (2000h:333).

The modern state can be understood, in part, as the development of the pastoral power of the Church. A number of changes constituted the newness of this pastoral power as it transformed into the modern state, including a transformation in its objectives. 'It was a question no longer of leading people to their salvation in the next world but, rather, ensuring it in *this* world' (2000h:334, my emphasis). In this context the meaning of the word 'salvation' is transformed, to become, instead, health, well-being, sufficient wealth, standard of living, security. 'A series of "worldly" aims took the place of the religious aims of the traditional pastorate' (Ibid.). The power of the pastorate increased and stretched far beyond the declining institutions of the Church, whilst the aims and agents of pastoral power split into two specialisms, one concerning the knowledge of man as individual and the other as population. Foucault redescribes the emergence of the modern state as the spreading out of pastoral-type power into the whole social body (Foucault, 2000h:334–5; 2007).

Some three centuries, a classical episteme and a form of power centring upon statecraft, separate the Reformation from the development of biology as a discipline and the inauguration of biopolitics proper. So we should be careful about overstating this case. Nonetheless, Foucault does describe the 'double bind' of power in which we are still tied up as this 'new pastoral power' (Foucault, 2000h:336), and he dedicates the larger part of the lecture series that forms his genealogy of biopolitical governmentality to the power of the pastorate (Foucault, 2007). It is, therefore, legitimate to propose that for Foucault the 'biological life' which is the object of biopolitical governmentality is something like the 'immanentised', worldly figuration of the eternal life of the soul, which was the *telos* of the Christian pastor. Foucault certainly wants to emphasise the continuity of pastoral power as *care* and as *individualisation* into the biopolitics of the present.

Foucault's focus on the pastor in the genealogy of biopolitics resonates with the final chapter of Arendt's *The Human Condition* in which she claims that it was specifically life (rather than work or action) that was able to take on such immense value in modernity because the event of modernity (the reversal of the *vita contemplativa* and the *vita activa*) took place in a specifically Christian context (Arendt, 1998:313–30). The good news that Christianity brought to the ancient world was that of the immortality of the human soul. This 'promoted the most mortal thing, human life, to the position of immortality, which up to then the cosmos had held' (Arendt, 1998:314). Life on Earth – the life that begins with birth and ends with death – takes on a new and immense importance in Christian thinking (an import it never could have held in Classical philosophy) because whilst '[l]ife on earth may be only the first and most miserable stage of the eternal, it still is life, and without this life that will be terminated in death, there cannot be eternal life' (Arendt, 1998:316). It is Christian thinking that first places a unique and, as it were, transcendent, eternal value upon mortal life.

The elevation of organic life to ultimate value in modernity is, according to Arendt, what you get when you cross the Christian elevation of mortal life to eternity and divine value with the modern rejection of divinity, collapse of eternity and demotion of contemplation. Life has asserted itself as the highest good of modern society and as the ultimate point of reference because 'the modern reversal operated within the fabric of a Christian society whose fundamental belief in the sacredness of life has survived, and has even remained completely unshaken by, secularisation and the general decline of the Christian faith' (Arendt, 1998:314).

Foucault emphasises the caring work of the pastor and Christianity as a form of power whilst Arendt focuses upon a philosophical and theological history of values, and upon Christianity as a formation of ethics. In both accounts, however, biological/organic life appears as the ultimate value or *telos* in modernity and in both accounts it appears as something like an immanentised version of the eternal life of the soul that is the subject of pastoral power and the *telos* of Christian subjectification. This rooting of biopolitics in the traditions of Christian care for life is evidence enough to insist upon the centrality of ethics and positivity to the vision of the biopolitical that Foucault and Arendt hold in common.

Normalisation, totality and individuality – the difference between Arendt and Foucault

There are, then, numerous resonances between Foucault and Arendt in the analysis of biopolitics. Both describe a positivity – a production of force, embodiment and historicity – in the productions of biopolitics and both identify a continuity between the Christian pursuit of pastoral care for the soul and the biopolitical valorisation and pursuit of life. This means that Foucault and Arendt are in strongest agreement upon precisely those issues that Agamben obscures in their writing. In fact, it is where Arendt most closely resembles Agamben that she and Foucault come into the greatest tension. The big point of disagreement between Arendt and Foucault concerns the character of normalisation and, specifically, the place of individuation within it.

A crucial issue upon which Foucault and Arendt clearly do not agree concerns the place of totalisation and individuality in relation to the normalising impetus that, for both, is associated with modernity. For Arendt, modern normalising society is totalising, mediocratising and excludes individuality (Arendt, 1998:40–2). For Foucault, modern normalising power is both totalising *and individuating*: normalisation works through processes of individualisation such as 'the case' (Foucault, 2000h; Osborne, 2008:110–111). As Frederick Dolan has argued, whereas 'Arendt sees normalisation as the result of anonymous, informal social pressure to conform, Foucault understands normalisation to proceed in a manner that is to a considerable extent "agonistic"'; for Foucault normalising power is addressed to citizens who are at liberty (Dolan, 2005:375). Modern bureaucratic organisation is totalising for Arendt whilst it is both totalising and individualising for Foucault. The difference between them on this matter is perhaps most starkly apparent in the contrast between Arendt's assertion that the 'no-one' by which modern society is ruled 'does not cease to rule for having lost its personality' with Foucault's insistence that we need, precisely, to lop off the king's head in our thinking and stop imagining power as something that is exercised by a personal sovereign (Arendt, 1998:40; Foucault, 2000a:122).

Further, Arendt argues that the despotism of the family becomes the model of government in the modern age. There has, she claims, emerged a new entity, 'society', and it constitutes a kind of massification of the family. The modern age, the age of society, is normalising because it is like a family, characterised by despotism and conformity rather than individuality and action. Action is excluded as the cost of behaviour, individuality is excluded from the 'public' (that is no longer public) and an immense force of conformism is exerted over all. The equality of members of modern societies 'resembles nothing so much as the equality of household members before the despotic household head' (Arendt, 1998 40). Foucault, in what could be read as an implicit response to Arendt's thesis, argues, in contrast, that with the emergence of the population the family disappears as a model of government, becoming instead its privileged *instrument* (Foucault, 2007:104–5). Not 'family style totalising despotism' but individualising, regularising, even autonomising and definitely dispersed *governmentality* characterises modern society, according to Foucault – normalising impetus and all.

The relationship between individuality, totalisation and modern normalisation (or normation) is, thus, considerably more complicated in Foucault's assessment than it is in Arendt's. In Foucault's account individuation (and subjectification) constitutes something like an internalisation or enfolding of community and its authority (see Dean, 1996; Deleuze, 1988:78–101), such that individuation is an agent, not opponent of normalisation. This difference between Foucault's and Arendt's thinking is further manifest, in inverse fashion, in their aspirational accounts of possibilities of genuine autonomy as politics (Arendt) and ethics/aesthetics of existence (Foucault). Whilst it might appear that Arendt's politics is collectivist and performative whilst Foucault's ethics marks a retreat to the self and is intellectualist, it is in fact the case that *both* Arendt's politics and Foucault's ethics are collective and take place in a public space. The difference is that the work of Foucault's ethics takes place within and upon a kind of internalised public space constituted in discourse and subjectification (Bell, 1996:93–4).

The space that this tension opens up between Foucault and Arendt is fascinating and raises all sorts of issues that can contribute immensely to our comprehension of both thinkers. Sadly the exploration of those issues is well beyond our present scope. What is important to note, for the specific purposes of this chapter, is that it is here, on the issue of the totalising nature of modern society, where Arendt comes closest in her thinking to Agamben's dismal account of the modern excepted state of affairs, that Arendt is *at odds* with Foucault. Arendt's *dis*agreement with Foucault marks precisely the point of her agreement with Agamben. We can, as such, hold on to the illuminating intersection between Arendt's and Foucault's thinking whilst at the same time rejecting Agamben's account of their ideas.

Conclusion

I have a great deal of sympathy for Agamben's apparent project: to point at the dead and to keep on pointing. The argument of this chapter is not intended as some kind of defence of biopolitics. Sadly, biopolitics doubtless is exactly as bad as Agamben would have it for all those who are caught in the thanato-political logistics of biological-type relations, whose death – political or physical – has been granted the role of vitali-sation of the life of the population by the casting agents of contempo-rary neo-liberal, neo-imperial and neo-theocratic biopolitics. The point of positive critique is not to be cheerful about the world. It is to be realistic – to describe processes in the world in terms of their real posi-tivity, their impact, their expansion of force.

Whilst Agamben's effort to point to the dead and to denounce has an important place in moral critique I fear that his rewriting of Foucault and Arendt on biopolitics risks reversing crucial advances in our abil-ity to understand and undermine the operation of oppressive power. It seems to forget that biopolitical, as all political, discourse has to work performatively, that the power has to find a way to get a 'hold', that it must appeal to its audience. Foucault and Arendt are part of the move-ment towards the comprehension of that appeal, of the reason for the applause. And if we lose this dimension of their thinking we lose the major part of their contribution to political theory and sociology.

Agamben equates biological politics with a force of reduction, objec-tification, domination, elimination and order. We must assume either that such a politics has no experiential 'hold', no economy of experi-ence making it appealing or necessary, or that fixation, domination, reduction, objectification and elimination are appealing, attractive, or at least incorporating experiences for people. Neither seems an ade-quate position. Certainly neither is in keeping with either Foucault's or Arendt's accounts of political discourse.

Philosophers and political theorists of the left are always too ready to carry the reaction of horror, the need to denounce, into the *ontology* of the social. Too ready to adopt the dual position of denouncing power and exonerating victims; to shout about those individual lives which are caught up in the terrible machines, about whom 'nothing strikes us more powerfully than their innocence' (Arendt, 1968:6). The task of the prophet is not, however, the task of the political analyst and strategist, nor of the *ethical* thinker for that matter.[7] The moral denunciation is not the fact of the case. To seek the innocence of victims is to subordi-nate the ontology of politics to the moral endeavour of finding a place outside of power from which to denounce it. It is, in the end, to grasp at *impotence*. The reaction of horror – not the action of revolt.

Feminism, post-colonialism and thus post-structuralism have hap-pened, and many lessons have been learnt in the attempts to grasp action, not the least of which is the one about the need for realism. Great political and ethical effort has been expended in the endeavour to learn how to 'take up the tools where they lie', where the very taking up begins first of all with the admission of such tools' existence (Butler, 1999:145) – the comprehension of the positivity of power.

Foucault's theorisation of modern power as biopolitical, along with Arendt's anthropologies of modern politics and conditions of humanity, constitute powerful contributions to those efforts – distinctive contributions that socialise Nietzsche rather than Freud. They demonstrate the centrality of biological values – health, vitality, process – and biological concepts to modern political discourse *across* the spectrums. In this they make abhorrent political rationalities and strategies more comprehensible – more addressable – rather than simply more abhorred. Further, they make strange these most familiar values, 'our' ultimate values – such as life, freedom, creativity and social security – demonstrating their historicity and consequent fragility. As such they make us more free from ourselves, ironically, more alive, opening spaces and fractures in which to imagine and embody alternative real politics. If Agamben writes the positivity – the care, the ethics, the intensity, the processuality, the expansion of forces – *the experience* – out of biopolitics then he destroys those efforts. Such a move can do nothing but return us to a hole where we would wallow in innocence and impotence, screaming into a void beyond power, whilst the biopolitical dreamers continue their play to the crowds unabashed.

4
'Post-Population' or 'Cultural' Biopolitics? Rethinking Foucault's Concepts Today, against Nikolas Rose

Nikolas Rose has been highly influential in the Anglophone social-science reception of Michel Foucault's thought and beyond. In particular he, alongside others such as Colin Gordon (Burchill et al., 1991), is associated with the popularisation of Foucault's thinking on governmentality and with the establishment of 'governmentality studies' as a sub-discipline of sociology, geography, political science and socio-legal studies – a sub-discipline focused upon the 'conduct of conduct' or governance through freedom (Rose, 1989; 1999; Rose et al., 2006). Much of Rose's work is addressed to the political history of medical knowledge and expertise, especially with respect to the 'psy' disciplines. In 1989 he argued that the proliferation of the 'psy' disciplines has been integral to the establishment of modern governmentality (Rose, 1989). In the past decade he has been concentrating on post-molecular transformations of biological knowledge, exploring the consequences for governmentality of recent developments in genomics, neuroscience, pharmacology and psychopharmacology (Rose, 2001; Rose, 2007). In this recent work Rose emphasises the plasticity and contingency of the conception of biological life that is at play in molecular biology, which emerged in the latter half of the twentieth century. He describes the efforts and intentions of scientists and patients to manipulate biological life at a molecular level in the name of individual health. The normativity that is at play in contemporary biology is, he maintains, a kind of somatic ethical work upon individuals. The norm of biological science is no longer population life (2001:13).

According to Rose, and Paul Rabinow with whom he penned 'Thoughts on the Concept of Biopower Today', the post-molecular transformations in the normativity and objects of biological science and medicine mean that bio*politics* also can no longer be understood in terms of population life. Molecular biology is, they suggest, part of a general movement of power away from a concern with populations and pre-given communities towards a liberal capitalist concern with transformation and ethical medicalised somatic manipulation. 'Biopolitics' Rose proclaims 'now addresses human existence at the molecular level: it is waged about molecules, amongst molecules, and where the molecules themselves as at stake' (2001:17). Biopolitics no longer pertains to the life of populations but, instead, to 'the politics of life itself' (2001, 2007). These conclusions, about the non-population-centred character of contemporary biopolitics, have consequences for how we should interpret and apply Foucault's ideas today according to Rose. In particular he suggests that the analysis of biopolitics is not relevant to contemporary race politics and racism; that specifically *biopolitical* racism is a thing of the past (Rose, 2001:2–7; Rose & Rabinow, 2003:17–21). Whilst Rose certainly remains critical of the practices and economies of biological science, pointing to their integration with the production of biocapital, there is a markedly optimistic tone in his account. The (supposed) introduction of contingency, responsibility and choice to biological life has generated a kind of vitalisation of politico-ethical as well as biological life; 'a spiritualization of the flesh, [a] sensualisation of ethics' (2007:258). This does not only open doors to the production of biocapital (and the attendant normativity of bio-ethicists) but also to a new ethics 'one that is embodied in the judgements individuals make of their actual and potential choices, decisions, and actions as they negotiate their way through the practices of contemporary biomedicine' (Ibid.:8).

In this chapter I will problematise aspects of Rose's interpretation of Foucault's theory of biopolitics and its applicability today. In particular I will problematise the way that he seemingly elides biopolitics with all, any and only politics of the somatic. Against Rose I will insist on the continuing importance of distinguishing between biopolitics and disciple as we talk about biopolitics today. Further I will challenge his contention that the economies of population life are no longer in operation, suggesting that population life may have been refigured in cultural and economic terms rather than having disappeared in the era since the molecular revolution in biological science. In particular I will suggest that (specifically dynamic, inclusive) biopolitical racism, and its dangers, might operate today in the experiential economy of culturalist politics. Again, the objective is not to negate, disprove or dismiss an account of the biopolitical. Rose's writings on medical power-knowledge and powers of freedom are a great contribution to the genealogy of the present. The objective remains positive-critique in the sense of pluralizing and adding to our accounts of reality. The problematisation of Rose's position can be seen as an attempt to recover distinctions and descriptions in the analytic of biopolitics that are in danger of becoming lost – a negation of negation. Following this I will move on to the more positive-critical work of building alternative, additional descriptions of post-molecular biopolitics. I will draw some highly speculative pointers towards alternative histories of the transformations of biopolitics since the Second World War suggesting that non-somatic formations of the trans – economics and education – might have since become primary in the in-corporeal-isation, the embodiment, of population life.

An alternative perspective on the application of the concept biopolitics today, and on the transformed fate of biopolitics in the period since the Second World War, is that of Michal Hardt and Antonio Negri, whose book *Empire* and its sequel *Multitude* have been highly influential, both amongst social scientists and a wider activist audience (2000; 2005). Hardt and Negri incorporate the ideas of biopolitics and biopower into an autonoma neo-Marxist account of global politics. For them biopolitics is not contemporary with capitalism but rather describes a second-stage of capitalism, wherein global biopower and a decentralised Empire have taken the place of capital and the bourgeoisie, whilst the immaterial, autonomous multitude has taken the place of the proletariat and its labour power. Hardt and Negri maintain that nineteenth century capitalism was disciplinary, with power taking hold of and exploiting the bodies of the workers. For them 'biopolitics' indicates a new depth of exploitation whereby power takes hold of and invests the very life, the vitality, creativity and immateriality, of bodies. In line with the tradition of Marxist analysis, this totalisation of power – taking hold of life, exploiting surplus life – is at the same time a virtual emancipation of life from power. The power of the multitude, that is developed and exploited by Empire/capital, is specifically immaterial, affective, informational and creative, labour and it does not depend upon capital for its development, deployment or transmission. As such, according to Hardt and Negri, the (vital) power of the multitude is radically more autonomous from, and able to escape, capital than was that of nineteenth-century disciplined labouring bodies.

Although Hardt and Negri seem to be talking about the positivity of biopolitics, their epochalising approach is so far removed from the perspective that is being developed in this study that it is appropriate to set them aside (especially given the effort to engage in *positive* critique). Indeed, their interpretation of Foucault's concept of biopolitics owes more to Gilles Deleuze's comments on Foucault (especially his 'post-script on control societies') than it does to Foucault's own writings and lectures (Deleuze, 1995). I agree with Rose that the totalising character of Hardt and Negri's interpretation of biopolitics and biopower (as the power of global Empire, the power of capitalism) robs the concept of its analytical purchase (Rose & Rabinow, 2003:7). Moreover their iden-tification of biopolitics with a post-disciplinary and post-nineteenth-century new epoch of power obscures the entire historical context in relation to which Foucault developed his ideas. As we have seen, Foucault identifies the emergence of biopolitics with the *beginning of the nineteenth century*. It is associated with the development, not the (supposed) decline, of the nation-state. The biopolitical era began with the nineteenth century and it is not disciplinary but regularising. Also Foucault insists that neither discipline nor biopolitics are capable of tak-ing hold of the entirety of society. As we have seen, they are addressed to different domains – to different levels – and thus even though their values and objectives conflict they can act in concert and agreement. Certainly biopolitics cannot supplant discipline, for it could not oper-ate unless there were also disciplinary institutions. If there ever was a non-biopolitical 'disciplinary society' then we would have to locate it in the seventeenth and eighteenth centuries – the time of natural history (Foucault, 1970) and police craft (Foucault, 2007) – before the technolo-gies of biopolitics and biology had emerged, not in the nineteenth and early twentieth centuries as do Hardt and Negri. Given the inability of disciplinary institutions to take hold of mass phenomena, however, the term 'disciplinary society' seems far too strong even here.

Whilst Hardt and Negri do, in a sense, refer to the positivity of bio-politics their approach is antithetical to that developed in this book. For them there is a positivity to biopolitics precisely because it makes work-ers, freedom and life better able to *escape* the confining and exploitative power of capital and Empire. In contrast I am interested in the positiv-ity that is immanent to biopolitical power – that is an aspect of bio-political rationalities and technologies even at their most exploitative and oppressive moments. I am arguing that life, capacity, contingency, creativity, experience and empowerment are intrinsic to and produced through biopolitics or biopower, they are not an alternative. Life can-not escape biopolitical power: biopolitical power/knowledge and 'bio-mentality' constitute the structures of experience in which creative life emerges and is perpetually produced. Rose, who posits the vitalisation of and care for bodies as immanent to the extension and exercise of biopolitical governmentality, and contingency as central to at least contemporary biological knowledge, is much closer to this position. Indeed, Rose's work on governmentality has been a significant influence for me. A critical engagement with his ideas on the utilisations of Foucault's concepts today is, as such, a far more 'positive' challenge, and far more interesting, for the project of this book.

Rose on post-population biopolitics

As noted in Chapter 1 the latter half of the twentieth century saw the previously disparate branches of biology, physiological bio-chemistry and evolutionary genetics, come together spawning various new branches of study that operate at a molecular level (Jacob, 1973:299). In the past decade Rose has been investigating the consequences of developments in some of these new molecular biologies; genomics, neuroscience, pharmacology and psychopharmacology, exploring their impact upon questions of social control, mental health and racial politics (Rose, 2007). He argues that the truth regime of biopolitics radically transformed in the second half of the twentieth century and that a new 'molecular biopolitics' has emerged that is entirely different from the eugenicist biopolitics of population life that Foucault described (2001:1). Molecular biopolitics, according to Rose, is not concerned with population (2007:58).

This new biopolitics is addressed to the plasticity of the biological and to the control of risk, rather than to the regulation of and care for the life of the population taken *en mass*. In the second half of the twentieth century, he argues the 'norm of individual health replaced that of the quality of the population' (Ibid.:13). He emphases the belief in, and desire for, the capacity to *transform* biological destiny at a molecular level, on the part of contemporary scientists and patients, in the name of individual health, and he contrasts this contemporary 'molecular' 'ethopolitical' 'risk control' biopolitics with an older expert imagining of life 'as an unalterable fixed endowment, biology as destiny' and its attendant biopolitics wherein 'the reproduction of individuals with a defective constitution [was] to be administered by experts in the interests of the future of the population' (Ibid.:20–1). He maintains that biology is no longer a science of determinations and that biopolitics is no longer a politics of population life. Biology operates at another scale (at the molecular) and that scale is full of contingency and hope. '[O]ur somatic, corporeal neurochemical individuality has become opened up to choice, prudence, and responsibility, to experimentation, to contestation, and so to a politics of life itself' (2007:8).

On the continued relevance of distinctions: biopolitics, anatomo-politics and the molecular

Rose collapsing biopolitics, discipline and molecular biology together

Despite Rose's frequent insistence upon the importance of historical specificity in Foucault's concepts (e.g., Rose & Rabinow, 2003:7), he happily refers to the new formations of genetic responsibilisation and molecular ethics as 'biopolitics' – divorcing the term from Foucault's definition and dismissing the argument that the bio, the life, of biopolitics is only produced in the (limit-) experience of population (a point that was explained and defended in Chapter 1). It seems, then, that Rose is diminishing the specificity of Foucault's terms and equating 'biopolitics' with something more general and ahistorical such as 'the politics of the somatic' or, perhaps, 'the politics of medical expertise'. In the same vein Rose, alongside Rabinow, treats the distinction between disciplinary anatomo-politics and the biopolitical politics of population life as irrelevant to the analytics of biopolitics today. Rose and Rabinow's article makes no mention of this distinction and neither anatomo-politics, nor discipline, appear in the index of Rose's book on contemporary biopolitics (2007). Discipline and anatomo-politics are implicitly subsumed into the biopolitical. This amalgamating approach to the different formations and politics of embodiment reflects Rabinow's interpretation of the coming together of physiology and genetics that François Jacob described (Jacob, 1973:299). Rabinow maintains that since this juncture the two discrete modes of address to the body that Foucault delineated 'are being rearticulated into what could be called a postdisciplinary, if still modern, rationality' (1999:407).[1]

In a move that seems designed to substitute for the distinction between discipline and biopolitics, Rose and Rabinow *do* talk about the distinction between the 'micro' or 'molecular' and the 'macro' or 'molar', borrowing from Deleuze and Guattari's vocabulary. They equate the molecular with 'individuation' and the molar with 'collective politics' and go on to suggest that where the molar was privileged in the era of eugenics the molecular is privileged in the contemporary era of liberal capitalism (2003:15–16). In fact Rose and Rabinow's appropriation of these terms to (presumably) stand in for the anatomo- and bio- political seems ill judged. When Deleuze and Guattari use the terms molecular and molar they are not referring to a difference in scale but rather to a difference in the modes of differentiation, organisation and formations of embodiment. The molecular is qualitative and transforming, it is the more vital or creative, the nomadic. The molar is the quantitative, the centralising, the controlled, the statist (Deleuze & Guattari, 1988:35). If this is what Rose and Rabinow understand by the terms then their deployment really does obliterate Foucault's distinction between discipline and biopolitics altogether, for in the scenarios that Foucault describes it is precisely the more individuating and micro practices of discipline that are the more 'molar' in Deleuze and Guattari's terms – they are more centralising, quantitative, statist and reductive – and it is biopolitics that pertains more to the 'molecular' realms of the unconscious, creativity and affect. I do not wish to claim that the molecular maps on to the biopolitical or the disciplinary onto the molar, but simply that if one *were* going to draw the association in a manner that is in keeping with both Foucault's understanding of biopolitics/discipline and Deleuze and Guattari's understanding of molecular/molar, it should in fact be in the opposite alignment to the one which Rose and Rabinow seem to suggest. In making the molar and molecular substitute for the biopolitical and the disciplinary Rose and Rabinow in fact wholly obscure the meaning of both sets of concepts.

Recovering the distinction between discipline and biopolitics

Rose's disregard for the distinctions between biopolitics and anatomo-politics (and, I would say, molecular-ethopolitics) is highly problematic from the point of view of this book which maintains that biopolitics and discipline refer to different and often contesting structures of experience. The reconciliation between biochemists and geneticists through the deployment of scale hardly amounts to a reconciliation between two fundamentally divergent formations of experience; of temporality, spacing and normativity. Whilst it is without doubt correct that new formations of power, knowledge and embodiment have been carved out in the post-molecular context – the era of the superfold and super-man, as Deleuze describes it (1988:107–10) – it does not follow that the distinction between biopolitics and discipline has thereby become irrelevant, as Rose and Rabinow have concluded. In order to maintain an account of biopolitics as the economy of modern *experience* that is not reducible to that of modern scientific expertise, it is crucial to maintain this distinction. There are three major problems with the assumption that the molecularisation of biology has made the distinction between biopolitics and discipline redundant for the discussion of the relevance of Foucault's ideas on biopolitics today.

The *first* and most obvious problem is that, whether or not the two forms of power persist *today*, the distinction remains relevant to understanding recent history and, most certainly, to understanding the sense and utility of Foucault's concepts. Indeed, if the politics of life no longer stands in contrast to an individuating, rationalist, disciplinary anatomo-politics, and if it is *not* a politics of population, then it must in fact be something very different from the formations that Foucault wrote about in the context of 'biopolitics'. If we do want to maintain that the politics of life no longer pertains to the population, having turned molecular, then we should find another term for the post-molecular politics of life, and retain 'biopolitics and discipline' as historical concepts, both of which are relevant to the analysis of nineteenth- and early-twentieth-century European states and resistance movements. Doing away with the distinction whilst persisting with the use of one term (biopolitics), as Rose and Rabinow seem to propose, feeds the misconception that pre-molecular rationalities of politics, including such practices as eugenics, can be correctly characterised as disciplinary and *not* biopolitical formations.

Second, the production of new formations of power/knowledge does not in fact necessitate the dissolution of the old. Foucault's own positive, additative, approach to history assumes that this is not the case. The production of discipline did not mean the end of sovereignty or the law anymore than biopolitics meant the end of discipline and institutions. In line with this we should not assume that the production of new molecular and recombinant formations of embodiment has necessarily meant the demise and irrelevance of disciplinary and (properly speaking) biopolitical formations. Treating formations of power in the manner of successive eras is to participate in a singular view of power, the philosophical tendency (that had failed to chop off the king's head) of which Foucault was so critical (Foucault, 1978:89; 2000a:89). If we accept that, as Foucault argues, there is a genuine plurality of formations of power and embodiment, and if we see history as a heterogeneous series of events, productions and additions, then the fact of the emergence of molecular biology seems a slim basis for the assumption that discipline and biopolitics, as distinct formations, no longer exist.

This brings us to the *third* problem with Rose and Rabinow's treatment of the distinction between biopolitics and discipline as irrelevant for contemporary analysis; it seemingly equates biopolitics with biological science and medicine in particular. If biopolitics basically *is* medicine and the application of biological science, then the molecularisation of biology would indeed mean a molecularisation of biopolitics. But this does not follow from Foucault's analysis; if biopolitics were simply medicine and applied biology then the term itself would be superfluous. Foucault's work on biopolitics illuminates the uses of biological rationality and biopolitical formations of embodiment far beyond the context of applications of biological science and medicine. He states (in reference to the phenomenon of biopolitical racism):

> evolutionism, understood in a broad sense…became within a few years during the nineteenth century not simply a way of transcribing a political discourse into biological terms, and not simply a way of dressing up a political discourse in scientific clothing, but a real way of thinking about the relations between colonization, the necessity for wars, criminality, the phenomena of madness and mental illness, the history of societies with their different classes and so on. (2003b:256–7)

The development of biological rationality facilitated the production of a host of *values,* and a network of analogies and strategies for organising bodies and forces, that far exceed the scope of somatic well-being or the application of scientific expertise. Population is the embodiment of the nation, the civilisation, the class, or the empire, as much as it is that of the race or the species. The health and vitality of the race and the species, let alone of the other embodiments, have since they were carved out been conceptualised, performed, expressed and experienced in cultural, ideational, aesthetic and economic terms as well as in the somatic. The vitality of these bodies, the health and security of the processes of population life that they designate, was never contained within the scope of medicine or the management of soma – immensely significant as these have been.

Rose's treatment of biopolitics as a historical, experiential and political event is overly 'soma-centric' and this leads to an unfortunate conflation of the emergence of new formations of biological expertise and somatic practice with the obliteration of older and additional technologies of power, knowledge and embodiment.

On the continued relevance of 'biopolitical racism'

Rose on the irrelevance of the analytics of biopolitics to contemporary race politics

Further to his claim that contemporary biopolitics is not concerned with population life, Rose rejects the idea that the concept of biopolitical racism is relevant for understanding race politics in the present (Rose, 2001:6; 2007:184–6; Rose & Rabinow, 2003:17–19). Rose makes this claim as part of his analysis of recent developments in genomic medicine, particularly the 'revival' of the idea that racial difference has a basis in biological fact, now figured as genetic difference. He wants to counter the claims of those who suggest that the recent moves to consider race and ethnicity in genomic medicine 'mark a potential shift toward a racialised medical practice, presage the reawakening of a dangerous racial science, and represent a further turn in "genetic reductionism"' (2007:156). Such fears are, Rose suggests, based on a miscomprehension of the normativity and objects of genomic medicine, ignoring the fact that contemporary medicine is directed at the health of individuals not populations and that the identification of genetic difference in the present constitutes the creation of new ground for intervention and transformation, not a statement of determination. He writes:

> [W]e need to locate the current debates over race and genomics firmly within the transformed biopolitics of the twenty-first century. This is a biopolitics organized around the principle of fostering individual life, not of eliminating those that threaten the quality of populations... it is a biopolitics that does not seek to legitimate inequality but to intervene upon its consequences. Crucially it is a biopolitics in which references to the biological do not signify fatalism but are part of the economy of hope that characterizes contemporary medicine. (2007:167)

Rose accepts that race persists in the present political landscape, functioning as a mark of discrimination, a mode of identification, and basis of rights claims. He and Rabinow draw attention to the 'murderous racist wars that spread across Europe in the wake of the demise of the Soviet empire' as well as to persisting racial discrimination in the US (2003:18). However, they insist, 'appeals to racial identities... [since the mid- twentieth century] needed no justification in the truth discourse of biology' as scientific expertise (Ibid.). Race, they state, is no longer related to 'a biological substrate', it is 'de-naturalised' (Ibid.).

Rose associates the idea that developments in contemporary genomics spells the return of eliminative eugenics with Agamben's ahistorical and negative account of biopolitics (2001:3–5; Rose & Rabinow, 2003:8–9). For Agamben, like Zygmunt Bauman, a thanatopolitical politics of population purification lies at the very heart of modernity (Rose, 2007:56). Agamben's reductive account of the biopolitics fails to recognise the radical transformations that have taken place in the domain of scientific expertise as well as the positive, life-maximising, work of all biopolitics. Rose wants to dispel the idea that Foucault's analytics of biopolitical racism and the eliminative moment of biopolitics is relevant in the present as part of an attempt to recapture the ground for thinking biopolitics from Agamben and insist upon the plural, historical and positive character of biopolitics.

Recovering the critique of biopolitical racism

Whilst Rose is right to object to Agamben's assertion that all contemporary biopolitics is a thanatopolitical project to make life homogeneous (as well as to assumptions that any attention to race in genomic medicine spells the return of eugenic racism) the movement from this critique to the conclusion that Foucault's analysis of biopolitical racism does not apply to the contemporary politics of race is problematic. Rose's argument effectively accepts Agamben's limited interpretation of biopolitical racism rather than taking seriously Foucault's own, very different, account of the phenomenon.

Indeed, Rose's contention that biopolitical racism is a thing of the past appears to rest on an identification of biological race politics with a pre-formist ontology and affirmation of eternity and fixed order corresponding to an attempt to establish a homogenous population. As Rose would have it we are *today* in the context of a politics of life without the population or biopolitical racism because 'life, today, is not imagined as an unalterable endowment, biology as destiny' and because the somatic has 'become opened up to choice, prudence and responsibility, to experimentation, to contestation' (2001:20). This implies an image of biological racism and relationships in the past that runs counter to Foucault's (and Hannah Arendt's) own thinking on the subject. We have seen in the previous chapters that 'biomentality', biopolitical racism and the politics of population life are, for Foucault, concerned with processuality, historicity and transformation – not with pre-formation and the adulation of order or stasis (the kind of thinking that can be associated with natural-history, physiology or disciplinary power and which Agamben elides with 'biopolitics').

As such it is hard to accept Rose's apparent conclusion that *because* contemporary conceptions of race and biology are centred upon contingency *therefore* biopolitical racism is no longer a relevant category to the analysis of our present. Biopolitical racism and biological type relationships are quite so dangerous precisely because they are imagined within a contingent field of potential, transformation, affect, evolution and indeed hope. From the perspective of the account of biopolitics and biomentality that is being developed here, the changes that Rose alerts us to, as he describes a new plasticity of biological life, appear as a transformation in the *location* of contingency in the context of biological knowledge, not as the introduction of plasticity and transformation into biological knowledge or life. His assurances that the presence of contingency, hope and a concern for individual health mean that we are safe from biopolitical racism appear to be based upon a conflation of eugenics with discipline, which Foucault's account of biopolitical racism is apt to displace.

I do not take particular issue with Rose's assessment of the politics of race in contemporary genomic medicine. There is not room to do justice to this topic here but I support Rose's contention that the identification of different propensities to diseases and treatments along racial and other lines is not, at present, constituting the lines of population-fragmentation upon which a vitalising politics of elimination and discrimination operates. Where I suspect Rose gets the analysis wrong is in the suggestion that there is *no such* politics of population, fragmentation and (what Foucault calls) biological-type relations taking place in the present time.

Whilst population might, as Rose argues, have become irrelevant to a biological science that has turned molecular, it might remain paramount in the formation of the political embodiments that are regulated and protected in national and international state politics, or fought for in socialist, nationalist, Islamist and ecological movements. Population has been an immensely expedient, effective formation of embodiment from the perspective of the generation of experience, affect, values and political authority, and there is little evidence that it does not remain so. A world that was post-population would surely be post-nationalist. Far from witnessing a decline of nationalism, however, the supposedly post-population era has instead seen a proliferation of nationalist, or ethnic-tribalist (Bauman, 1995:243–56), movements, as well as of political religious movements that closely resemble them. At the same time, new collective embodiments have emerged such as that of the EU, imagined in terms of an exclusive European civilisation (Bunzl et al., 2007). At the level of established states the vital(ist) imperative to 'secure the nation' – to always be doing something to be securing, regulating, enhancing the life of the nation – seems as strong as ever, more than five decades after the molecular revolution in biology. When emphasis is placed upon biopolitics as a formation of experience, not simply as a politics of biological science, it is evident that the analytics of biopolitics might be relevant to the comprehension of cultural racism, culturalist nationalism and regionalism.

Rose makes biopolitics into something *much* more specific and much more Foucauldian than does Agamben. In his explanations of what biopolitics is about it is very clear that biopolitics pertains to the politics of a specific history of knowledge. It is also clear that the development of that knowledge has been an immensely 'positive', vitalising development, at least with respect to people's somatic wellbeing. I do not contest Rose's account of the politics of contemporary medicine and biological expertise. However I do think that his interpretation of 'biopolitics' is too narrow; too tightly bound with the politics of medicine and biological expertise. Rose seems to reduce biopolitics to *nothing but* medicine and the expertise of biological scientists. In Foucault, however, the emergence of biopolitics heralds a whole new regime of values – outshining aristocratic spectacle and analytics of blood, constituting new grids of historical intelligibility (Foucault, 2003b), orchestrating new originations of meaning-value in sexuality and life, mobilising entire populations, even rendering massacres vital (Foucault, 1978). The analytics of biopolitics can help us to think about many things other than the political uses and abuses of biological-scientific expertise and medicine, pertaining to questions of 'political spirituality' (see Foucault, 2000e:233). Foucault's theories of biopolitics speak to crucial questions concerning ethics and the politics of culture.

Rose radically overplays the significance of scientific authority to the force of biopolitics. For him biopolitics is all about the application of biological science and expertise to the political (broadly understood). The authority of claims in biopolitical discourse and practice is, then, assumed to be vested in the authority and epistemic capacities of biologists. Whilst I would not contest the idea that the authority of biological and medical experts has been important in the formation of modern governmentality I maintain that there is an additional dimension to authority in biopolitics as Foucault describes it, that of experience, embodiment and the production of vitalist values. The formations of experience and embodiment that biopolitics engenders are themselves powerful planes of affective investment, capable of producing grounds for political authority without *necessary* recourse to that of scientific expertise. Experience as life, not just the *scientific* knowledge of life, is utilised in the generation of epistemic and moral authority for modern regimes of biopolitical power. Population life itself can act as an immanent plane of authority and investment for biopolitical practice and values. The authority of biopolitics is not reducible to that of the scientist or of scientific knowledge. Biopolitics is *also* about the production of a specific collective trans-organic embodiment that engenders a particular, particularly vitalist, horizon of culture, visibility and sayability. This embodiment – the population – the unitary living plurality constituted through transmissions of reproduction and mechanisms of evolution, is in and of itself productive of affective reality, subjectivity and the (limit) experience of moving beyond finite singularity; it is productive of experience that engenders the possibility of authority, epistemic capacity, rhetorical and political appeal, and meaning-value. As argued in Chapter 2, it is the dynamic connection between bodies as capacities, not their somatic nature, that is crucial in the constitution of this trans-organic and limit-breaking embodiment. Those connections always did include flows of information, education and training in addition to those of blood and semen. The post-molecular age might be characterised not by a demise of the population but by a reconfiguration of *which* flows, and *which* capacities, are primary in its constitution such that blood and semen lose ground to education and economic integration.

In contrast with Rose's suggestion that we have seen an end to population biopolitics, I propose that what has in fact taken place is a coming to dominance of cultural and constructivist rubrics of population biopolitics. We will now turn to the consideration of anti-biologistic social-constructivism, which has been a defining aspect of dominant political discourses – both of the right and the left – in post-second world war Europe and beyond. Rather than a movement against the politics of population I will suggest that this 'anti-biologism' represented a widespread 'culturalisation' of population politics. Whereas Rose places developments in biological science at the fore in his analyses of the changing face of biopolitics my account will suggest that political struggle over the ontology of race is primary. Perhaps the rapid development of molecular biology in the post Second World War period is, at least in part, a *result* of political struggle concerning racial categories.

Social constructivism, cultural biopolitics and racism after Auschwitz

European and Northern American political discourse and social science in the post-Second World War period was characterised by a widespread 'anti-biologism'. In the early twentieth century, eugenic ideas had been immensely popular throughout Europe and the United States. After the Second World War and the defeat and demonisation of Fascism, however, there was a general rejection of eugenic ideas in non-fascist Europe and America. Eugenics was now associated with the Nazis and the most horrific practices of eliminative eugenics (Taylor Allen, 2000:449; Rose and Rabinow, 2003:17; Rose, 2007:167–8; Skinner, 2007:933–4; Ziegler, 2008:233–4).[2] Naturalism, biologism and 'scientific racism' came under fire in the post-war blame games across the political spectra. These wranglings issued in a widespread celebration of social-constructivism as a supposedly emancipatory and anti-discriminatory ontology.

Anti-racist social constructivism

The most obvious movement against biologism is that of anti-racism. Thinkers such as John Rex positioned themselves as experts and educators of the public charged with the task of overcoming the ignorance and prejudice of racism, largely by contesting biological accounts of racial difference. The potency of anti-biologism in the sociology of race remains strong. In 2007 David Skinner could convincingly claim that '[s]uggestions that an approach intentionally or unintentionally contains some residue of biology remain one of the most powerful and contentious ways of criticising sociological work on race and ethnicity' (2007:936). Arguably, a number of other anti-biology positions that have developed since the 1960s were motivated by the desire to challenge or gain distance from racism. For example, Vikki Bell argues that feminist 'anti-essentialism' was motivated by the problem of racism (Bell, 1999:114–16). A radical anti-biologism was articulated in post-war feminist thought, with biologism and Darwinism being widely characterised as an 'ideology of the status quo' (e.g. Rosenberg, 1975:142). This feminist anti-biologism will be discussed at length in Chapter 5.

Neo-liberal social constructivism

It is crucial to note that this strategy of attempting to gain distance from fascism through turning to, or affirming, constructivism was not at all a move restricted to the political left. Foucault details a parallel movement in the context of liberal political economy (Foucault, 2008:120–21).

The Ordo liberals, although writing before the Second World War, gained popularity in the post-war period, becoming a key voice in the reconstitution of the German state. The Ordo liberals argued that the rise of fascism could in part be attributed to the naïve naturalism of nineteenth century liberal philosophy; or rather, that the naïve naturalism of nineteenth-century liberalism was responsible for the weakness of liberalism, and the failure of the liberal state in Germany, which in

turn enabled the unchecked expansion of the state, monstrously manifest in the rise of the Nazis.

The naïveté of classical liberalism's naturalism (according to the Ordo
liberals) lay in the faith that the nature in question – the natural, self-
regulating behaviour of market forces – would come into being of its
own volition. It is this naïve naturalism that permitted the nineteenth-
century liberals to espouse the principle of *laisser-faire*, of establishing
markets as a domain *beyond* legitimate governance, to be left to their
own self-regulation. When you uphold the principle of *laisser faire*,
according to the Ordo liberals, you are thinking of the market as a sort
of given nature, 'something produced spontaneously which the state
must respect precisely in as much as it is a natural datum' (Foucault,
2008a:120). The Ordo liberals essentially agreed with classical liberalism that the market behaves as an auto-normative, natural field of
force (and site of verification-falsification for governmental practice
(see Ibid.:32)) – indeed the Ordo liberals wanted to radically extend the
domains of life to be regulated through the laws of the market. However
they maintained that for this nature, market behaviours and forces, to
emerge, the market conditions would first have to be *constructed*.

Competition, for the Ordo liberals, is not a universal instinct but
rather a formal structure that is given to intuition only in specific conditions; 'competition as an essential economic logic will only appear and
produce its effects under certain conditions which have to be carefully
and artificially constructed' (Foucault, 2008a:120). As such, neo-liberal
attention focused upon the ever-extending development of mechanisms
to produce markets and market behaviour. Whilst remaining *anti-state*,
neo-liberalism has been able to drive towards ever-increasing, deepening, extending governmentality; towards the continually increasing
construction of competition and market conditions.

Racist social constructivism or 'cultural racism'

The idea that new forms of 'cultural racism' have emerged since the
Second World War, forms that are compatible with variants of social-
constructivism, is widely held. The argument was spear headed by
theorists of race and identity who are particularly interested in the
power of culture, such as Stuart Hall (2000) and Paul Gilroy (1987) and
now receives broad support in the context of discussions surrounding
'Islamaphobia'. For example, when Matti Bunzl argued in 2007 that
Islamaphobia in contemporary Europe has to be seen in terms of a new
normative paradigm wherein at stake is no longer a biologically constituted racial nation, but a culturally constituted European civilisation,

his claim received support from a politically and philosophically diverse range of invited commentators – including Dan Diner, Brian Klug, Paul A. Silverstein, Adam Sutcliffe, Ester Benbassa and Susan Buck-Morss (Bunzl et al., 2007).

The idea that new cultural racisms can be seen as something like a refunctioning of older biological forms develops from the insight that the dynamic conceptions of collective life upon which culturalist discourses draw – conceptions such as civilisation and cultural development – have a shared genealogy with those of biological evolutionary racism in nineteenth-century European thinking and imperialist practice. Historian of anthropology George Stocking has, for example, demonstrated the influence of Darwinian and other evolutionary biology on the development of anthropological and social-scientific conceptions of culture and civilisation (1968), whilst a recent volume of collected works in political theory makes a wide ranging case for the continuity between contemporary and nineteenth century 'civillisational' imperialist discourse and politics (Duffield & Hewitt eds., 2009).

Arguably a shift in overtly racist discourses from biological to cultural racisms follows a shift in the authority concerning the knowledges of difference from biological sciences to cultural and social science. Sociologist of race science David Skinner has pointed out that the second half of the twentieth century saw a dramatic movement of authority whereby 'the baton for understanding race differences and managing race relations had passed from the natural to the social sciences' (2007:935). Anthropologist David Scott has suggested that a constructivist, relativist discourse of difference as cultural and linguistic has substituted for biologism as the ground for a 'post-ideology' inscription of western hegemony (2003).

If thinkers such as Bunzl are correct, then cultural racism is more pernicious, and more central to the mechanisms of governance in contemporary Europe, than is biological racism. Certainly a hostile and supremacist conceptualisation of Islamic culture is a defining political issue in the present. Bunzl cites comments from influential far right European political parties, such the remark from a leader of *Flemish Interest's* that 'Islam is now the No.1 enemy not only of Europe, but of the entire free world' (Filip Dewinter, cited in Bunzl, 2007:40). The English Defence League (EDL), created in 2009, is a violent protest group who describe themselves as a 'counter jihad movement' and who have been characterised by as 'the most significant far-right street movement in the UK since the National Front in the 1970s' by *The Guardian* (EDL, 2010; Taylor, 2010). They constitute the latest chapter in an important

Islamaphobic strand of political culture in Britain, seeking to provoke violent unrest and tension in a number of cities. Unlike the leaders of *Flemish Interest* the EDL publicly reject the term 'Islamaphobia'. However footage from a Guardian undercover investigation of hundreds of their members singing 'we all hate Muslims, we all hate Muslims' suggests the appropriateness of the term (Bunzl, 2007:40; EDL, 2010: Taylor, 2010). Countering the idea that such sentiments are only relevant to a radical fringe in Europe, Bunzl traces the movement by which far right parties from Austria and elsewhere have been able to shape the mainstream centre-right European political agenda, provoking such things as a move against support for Turkish accession to the EU on the part of centre right politicians such as German Chancellor Angela Merkel. The EDL add force to Bunzl's suggestions that Islamaphobia is primarily concerned with cultural, rather than biological, difference. Although there are reportedly organised racists and fascists in positions of influence within the movement, spokespeople for the group are adamant that it is a 'non-racist' organisation (Booth et al., 2009), by which they seem to mean that people of any colour are welcome in the movement. Their website states that 'we invite people of all races and faiths to join us in this campaign to awaken our sleeping Government to face up to and deal with the Jihad in our country, which threatens the very foundations of the freedoms won so dearly for us by past generations' (EDL, 2010). It is hard to believe that the pluralism about faith extends to welcoming Muslims. However footage from *The Guardian* investigation into the movement does demonstrates that in skin colour, ethnic origin and religion the group are far from monochrome (Taylor, 2010). Of course plural forms of racism intermingle and inform one another in this and other such contexts. With Bunzl and Rose I am attempting to indicate a shift in dominant emphasis from biological to cultural racism, not to suggest that biological racism is dead and buried.

The EDL and Flemish Interest might represent an extreme wing of cultural-racism but we can situate the shift in far right thinking from biology-centred, to culture-centred, racism, in the context of much broader shift in the imagining of difference, dynamics and fragmentation. A considerably more 'acceptable' manifestation of population life being presented as divided into dynamically related hierarchical fragments of culture in British political discourse, was former Labour Prime Minster Tony Blair's assertions that gun and knife crime, and a lack of respect, cannot be wholly explained in terms of factors such as economic inequality but must also be put down to a distinctive black culture (Potter, 2007). These comments resonate with the widespread condemnation of working class culture as 'chav culture' that is held to be responsible for social decline and 'anti-social' behaviour – legitimising moves towards tighter and tighter legal control upon behaviour. The discourse of 'chav culture' can be regarded as a form of 'social racism', as was argued by polemicist Julie Burchill in 2005.

Forms of cultural supremacism are also central to the productive imagining of international political and social order. Most obvious is the Imperialist discourse concerning the oppressive nature of 'non-democratic' cultures that has been utilised in the legitimisation of US led military incursions into the Middle East in the past decade. More intractable is the concept of 'human development', which is institutionalised in the UN 'human development index' and embodied in the practices of millions of providers and recipients of international governmental and non-governmental agency across the global south. As Mark Duffield makes plain, contemporary development discourse is intertwined with global biopolitical projects of securitisation and fragmentation of life (2007; see also Beard, 2006). Cultural supremacism is at work in the machines of development, creating and shaping capacities to transform a world understood as dynamic, contingent and (at least potentially) progressing. Not only does the discourse of development legitimise an international hierarchy of global actors and attendant processes of imperialism, it also engenders practices of subjectification wherein cultural racism is central to the day to day self-constitution of development workers as moral subjects and is internalised by many of the people that they work with (Kothari, 2006; Duffield, 2007; Lawson, 2007:35; Heron, 2007:88–99).

In both national and international contexts such normative descriptions of cultural difference do not simply give voice to prejudices or enable people to construct 'identities', they constitute specific organisations of visibilities, embodiment and capacities producing the possible paths of power and activities of governance. Discourses of cultural differentiation have accompanied a governance of life that works upon our cultural being, just as discourses of biological racism accompanied the formation of an array of eugenicist politics, directed at the improvement of population life through practices of care, control and elimination that operate at the point of biological reproduction or transmission of bodily fluids and touch. The new(er) cultural racisms describe alternative planes of contingency, animate new(er) possibilities of 'progressive' action, and valourise new(er) sets of activities as 'caring for life'. There is a focus upon practices of cultural formation and reproduction; educating, encouraging cultural change, capacity building and eliminating problematic culture.

Arguably the rationalities and, especially, the value systems of bio-politics and biologism have been reinscribed in a culturalist register – a register that is quite at home with the basic principles of social constructivism. If that is the case then the post-war critique of biologism may not have been successful in distancing 'progressive' thought and politics from the problems of modern racism and biopolitical rationality. Getting politics away from biological science has not necessarily meant an end to the experiential economy of population politics, bio-political racism or biological-type relationships.

If there has been a culturalist reinscription of population then the biopolitical economy of experience that Foucault analysed, including the investment of (biopolitical) subjects in (bio)political embodiments and the generation of biological-type relations, might persist in the post-molecular, post-biologistic context. Foucault's analysis of biopolitical experience draws the caricature of early twentieth century biologism as deterministic, pre-formist and conservative into question. If biological politics was invested in the experience of becoming, transcending limits, and transformation, and if biologistic ontologies apprehended the world as historic, contingent, vital forces, then some of the assumptions that are made in the 'anti-biologistic' celebration of social constructivism are problematic. At the least we can say that the affirmation of becoming and contingency are no necessary guard against the formation of biological-type relationships and racisms as Foucault has described them. Foucault's concept and analysis of biopolitics might be applicable to political discourses and rationalities today that refer to culturalist not biologistic ontologies.

Notes towards an alternative history of post-molecular biopolitics: economics, education and the embodiment of population life in post-war Britain

Foucault's theories of biopolitics can help us to think not only about the ever-expanding politics of medical practice, somatic citizenship and bio-technological consumption, as Rose implies. It can also help us to understand the experiential economy and positivity of non-biologistic population politics. With further research it would be possible to develop an account of a *culturalist biopolitics*, as paramount in post-molecular/post-Second World War European political discourse: an account pertaining to issues of experience, meaning and value not only with respect to the somatic, but also to cultural, educational, or formalist articulations and embodiments of population life. What follows are some highly speculative and schematic notes towards such a history in the context of post-war Britain. This is not intended as a substantive contribution to the political history of contemporary Britain, but as a continuation of our abstract exploration of the applicability of Foucault's concept of biopolitics today through the schematic indication of some of the areas to which that concept could be applied. In this section, I will set out some very schematic pointers towards an alternative history of post-molecular biopolitics in Britain. This is not an attempt to put forward a new empirical history of post-war British politics. It is simply an indication of a few of the fields to which Foucault's concept of biopolitics could productively be applied in the post-molecular context if we understand biopolitics in terms of an economy of *experience*, not just the politics of scientific expertise. These observations are not based upon substantive research but rather constitute notes towards possible future research into contemporary, or post-molecular, biopolitics.

Interpreting the separation of population and biology in terms of political struggle

I have suggested that biopolitics in the post-molecular period could be understood not in terms of a post-population biopolitics but rather as a situation wherein there has been a split between biopolitics and biological science. Rather than the amalgamation of disciplinary anatomo-politics with biopolitics, the molecular revolution in biological science and the movement of biology away from a focus upon population might be understood in terms of separation between biopolitical formations of knowledge and biological expertise. As Skinner suggests with respect to the specific discourse of racial difference (2007:935), social-scientists might have supplanted biologists as the authoritative experts on matters of population life.

Some might argue that such a split between biopolitics and biology would be of the nature of a time lag. Some would suppose that the politics of collective embodiment will sooner or later catch up with the natural sciences, giving up on the politics of population, as though culturalist formations of nationalism are a mere hangover from the eugenicist early twentieth century, and as though human science and political rhetoric simply follow the ontological lead of the 'natural' sciences. But, if we understand political struggle as primary in the formation of knowledge, then we *could* argue almost the reverse: that the apparent split between biological science and the biopolitics of the population was the *result*, of a political problem and of a solution championed by social scientists and politicians.

The political problem that might have initiated the split between biopolitics and biological science is that of how to persist with the nation state in biopolitical liberal democracies in the aftermath of the second world war, an aftermath characterised by horror and shock at the thanato-political totalitarian form of the nation state and at the intolerable trajectory that extremist eugenics had taken under Nazi rule. Arguably the solution to this problematic included the reconceptualisation of population as a primarily cultural, formal and economic body. In a sense the biological knowledge of race was positioned as, if not exactly the scapegoat (being undoubtedly massively responsible) then perhaps, the 'fall-guy' for all the woes of the biopolitics of population life. Arguably, the latter was enabled to continue in a revitalised constructivised form rinsed clean of (apparent) responsibility for the excesses of eugenics. As has been said, scientific racism, biological determinism and naïve naturalisms were widely denounced in the decades following the Second World War, as epitomised in the UN statements against scientific racism. The relative demise of 'scientific racism' and the progressive delegitimisation of 'biological determinism' was not, however, to result in the end of biopolitical experience or of the politics of population (which might have meant the end of the nation as a focus of affective power). Arguably population politics continues in the present whilst non-somatic expressions and manifestations of population life have come to play a more paramount role, partially substituting for the trans-organic corporeal connections of genetics and degeneracy.

Arguably the shift in political discourse and social science away from biologistic explanations and ontologies towards more constructivist 'sociological' understandings accompanied a reconfiguration of the biopolitical population. Foucault's analysis of biopolitical economies of experience and embodiment might be applicable to a welfare state population embodied though technologies of social insurance, articulating and investing capacities of caring, labouring and capital production. It might also be applicable to a post-welfareist neo-liberal population in which the capacities that are articulated are ideational and educational, even when they pertain to care and economic interaction.

Economics and education as alternative embodiments of population life in post-war Britain

In post-Second World War Britain, the new 'welfare state' would focus on social insurance and education as much as it would upon health. Social progress would be achieved through a fight against 'want, disease, ignorance, squalor and idleness' (Beveridge Report, 1942), rather than – as we might have expected a couple of decades before – degeneracy and racial decline. Whilst the importance of genetic genealogy was being downplayed, the trans-organic (or trans-individual, or trans-familial), character of wealth and education was brought into a sharp – newly institutionalised – relief. With the institutionalisation of 'social insurance' and the production of the welfare state following the Second World War a new complexity of relationships of economic interdependence were described and materialised. There are parallels between this welfare-statist embodiment of population life through technologies of social insurance and the embodiment of population life through sexuality in the nineteenth century as Foucault describes it. The technologies of social insurance connect the capacities of bodies, responsibilise bodies, define the 'truth' or true worth of subjects and set-up fragmentations, sub-groups, that are understood hierarchically in terms of those definitions.

Economic productivity as an aspect of population life was *complexified*, coming to include collective risk and responsibility, in addition to collective productivity and (national) success or failure. Mechanisms of comprehensive interdependence were established, giving greater body, greater corporeality, to national economy (or rather, perpetuating the corporeality of national economy that had already been well established in the context of total war). Economics is thus drawn deeper into the warp and weft of population life, substituting, to some extent, for the relatively weakened role of sexuality and genetic reproduction.

At the same time further education was established as a comprehensive right to be provided free to all with the Butler Act of 1944. In practice secondary education was made free and compulsory until the age of 15 in the year following the war. This was presented as a limited measure on the path to an ideal wherein the provision and compulsion would continue to the age of eighteen. The bigger transformation was the establishment of the principle that a liberal education was not simply a distinguished privilege but was, instead, an entitlement of all children and a practice that would serve as the *means* of social differentiation (Fraser, 2003:242). Rather than simply reflecting the privileges of birth education was to become the dynamic *ground* of social differentiation and dynamic relation. This was far from the universalisation of an existing right of the privileged. Education would be universalised but at the same time fragmented, theoretically split into a tripartite system of grammar, technical and secondary modern schools, but more factually split into an internally hierarchised triplet of public, grammar

and secondary modern. Thus we see a movement of fragmenting generalisation that echoes the generalisation of sexuality to the proletariat in
the nineteenth century, as Foucault describes it. The education system
is thus producing a unitary living plurality: dynamically related fragments of the population, the health of all of which is said to determine
that of the whole (the success of the nation). Membership of the different fragments was to be 'determined' by psychometric testing (itself
largely determined by class membership), rendering 'intellectual capacity' and capital crucial in the investment of individual bodies in and by
population life. Work, wealth and education would carry the agency of
individual bodies into the vitality of future generations and that of the
existent population-nation. Culture and education would oust genetics
as the principle fluid of reproduction.

A second-stage in the production of post-war welfare-state embodiment in the UK can be intentified with the establishment, since the
1980s, of neo-liberalism as the hegemonic political rationality. Neo-
liberalism can be seen as an alternative post-molecular regime of
biopolitical embodiment, one that is more formalist and more abstracting, moving further again from the explicit biologism and eugenicist discourses of early-twentieth-century biopolitics. Neo-liberalism
emerged, as we have seen above, in Germany, where the problematic of reconciling the biopolitical nation state with the horror at its
totalitarian potential and the excesses of eugenics was at its sharpest.
If we are correct in identifying the tensions around continuing the
nation-state in the aftermath of Auschwitz and extremist eugenics as
a cause for the development of formalist, educational, imaginings of
collective embodiment then it is no wonder that the 'constructivisa-
tion', formalisation or abstraction of the connectivity of population
life was more complete in the neo-liberal regime of power/knowledge
emanating from Germany. Economic relations, behaviours and vital
interdependence were *themselves* conceptualised, by the Ordo liberals,
as constructed *forms,* produced through education. The institutionalisation of neo-liberalism (heavily influenced by the Ordo liberals)
in the UK in the 1980s included a radical centralisation of education
production with the launch of the National Curriculum in 1988.[3]
Neo-liberalism considerably undermined the welfarist production of
collective embodiment in the guise of economic interdependence and
social insurance, but it placed even greater emphasis upon education,
culture and ideas (or forms) as the site of social reproduction, transmission and vitality. Neo-liberalism regards economic behaviour as
itself a matter of culture – as constructed forms of behaviour produced

through education. In turn it tends to view economic productivity as the measure of cultural life.

Conclusion

It is beyond the scope of this study to investigate this history empirically in a meaningful fashion. I am raising these pointers in order to suggest the types of areas to which Foucault's analytics of biopolitics *could* be applied, but I am not attempting to carry out that empirical application. *Very* schematically and *very* speculatively speaking we can suggest that, following the crisis in biopolitics brought on by the experience of extremist eugenics, total war, anti-colonial struggles and the ideological wranglings around Nazism, a split was introduced between the formulation of population and the concepts and mechanisms of evolutionary biology – between biopolitics (though not discipline) and biology. Following this split, cultural and economic manifestations of extra-somatic embodiment became (even) more significant than they had been in the first half of the twentieth century. It is possible to suggest that this event might even have had a part to play in the molecular turn in evolutionary biology, turning biologists away from some of the issues of population life (especially those pertaining to racial difference) and thus perhaps encouraging a focus upon individuation and the more deflationary approach to living beings of biochemistry, assisting in the development of the new recombinant molecular biological science. Either way, the politics of population life continues despite the relative decline of the authority of discourses concerning the genetic difference and determination of races.

As always we should avoid epochalising pictures of historical process. A move from a more somatic to a more culturalist biopolitics of population should be assumed to take the form of additional productions and emphases, not the move from one epoch of biopolitics to the next. Economic interdependence has always figured in the constitution of the modern population, whilst the idea of culture was one of the main concerns of nineteenth- and early-twentieth-century anthropology, studying the evolution and variation of Man as species. Conversely, the line between social and physical reproduction has been murky in post-war productions of population life and dynamic racism. Genetic thinking was never so broadly discredited in the context of class and especially sex difference as it was with respect to race. Also, post-biologistic liberal democracies continued to export eugenicist policies to so-called 'developing' countries, even where eugenics had been strongly denounced in a domestic context. If we are to note a shift between somatic and cultural biopolitics then it would pertain to a shift in emphasis and dominance, not in totalising epochs. If this approach is broadly correct then we should imagine that the molecular revolution and associated events have given shape to new forms, adding to and thus transforming the biopolitics of population but not displacing it. Foucault's analysis of biopolitics might be of great relevance to a number of phenomena today, phenomena that are not generally referred to as biological (nor treated as such by Rose).

These speculative notes on the biopolitical population in post-molecular Britain are not, as I say, intended as a substantive empirical account of that history – such an account would be well beyond the scope of the present work. My intention is simply to explore, in the abstract, the question of what Foucault's concept and analysis of biopolitics might help us to understand in a post-molecular context. Amongst the many areas upon which Foucault's analytics of biopolitics might shed light are the political appeal, the constraints, the necessary racisms and the perpetual intensifications involved in the experiential economy of welfarist and neo-liberal educational policy. In contrast to Rose who ties the history of biopolitics very closely to that of medicine and biological science I have suggested that alternative, culturalist embodiments of population are important for contemporary biopolitics and that Foucault's analysis of the economies of biopolitical experience might be applied to technologies of embodying population through social insurance and national education.

5
Eternally Becoming: Feminism, Race, Contingency and the Critique of Biopolitics

This chapter builds upon the interpretation of Foucault, biopolitical power and positive critique set out above by developing a specific example of biopolitical discourse and its critique – the example of feminism. Although Foucault does not himself discuss the issue, feminism is an exemplary site for thinking both the positivity and the critique of biopolitics. Twentieth-century feminism has been embroiled in the logics, values and problems of biopolitics – including the problems of biopolitical racism.

A genealogical account of feminism as a biopolitical discourse extends the arguments of this book, because it enables us to elaborate upon the positivity of biopolitics through the identification of its historical association with the evidently 'positive' emancipatory, empowering, transformative movement of feminism. This intersection of biopolitics and feminism illustrates the positive, productive and processual character of biopolitics – in line with Foucault's account. In the final lecture of the *Society Must Be Defended* series Foucault, then hero of the French left, upset his audience by pointing to the biopolitical nature of Socialism, implicitly suggesting parallels between Socialism and Fascism, explicitly describing Socialism as biopolitical and very often biopolitically racist. Of course there were many good reasons to raise such a critique of Socialism in the 1970s and it is characteristic of Foucault's reflexive ethics to raise problems that are uncomfortably close to home. But I think that this moment was also motivated by a kind of positive-critical impulse – an impulse to make biopolitics, racism and even fascism more comprehensible (which, of course, is not at all to say more justifiable) by pointing to the resonance between such values and those of his audience (and indeed himself). This chapter is motivated by a similar spirit. It is a discussion of both biopolitics and negative-critique in a context that is, for me, very close to home, precisely because it *is* close to home. I want to understand biopolitics and biopolitical racism, from 'the inside', to really get a grip on the biopolitical economy of desire. As such this chapter explores the biopolitics and negative critique in relation to feminism precisely because I do understand and strongly empathise with feminism.

Further, feminists have been amongst the most powerful critics of biologism and biological discourse, with a great range of attitudes to biology having been adopted by various feminists and feminisms. Thinking the critique of biopolitics or bio-mentality in relation to feminism does, therefore, provide an excellent opportunity to compare and contrast different ways of critically engaging with the biological. I will explain and defend the positive, Foucauldian, critique of biopolitics through contrast with the negating 'ideology critique' of biologism set out in the 1970s by materialist feminists. Whatever its many virtues, I will suggest, the anti-biologism that materialist feminism promoted may have done as much to mask and protect some despicable components of biopolitics as it did to confront others.

The genealogy of feminism and biopolitics is also significant because the relationship between contingency, vitality, biological embodiment, knowledge, racism and feminist politics persists as a significant problematic. Recognising the historical agreement between biopolitical racism, feminist emancipatory politics and ontologies grounded in contingency has consequences for how we think the possibilities of feminism in relation to (cultural) racism and 'supremacist-specificiation'[1] today. It draws into question the (often assumed) association between the assertion of contingency and the promotion of minority empowerment – biopolitical racism, as we have seen, is very much a politics of contingency, transformation and process.

The genealogy will commence with a discussion of 'early' feminism, arguing that this can be understood as one amongst other discourses and practices of modern biopolitics.[2] The particular issue of eugenicist feminism will be highlighted as a clear example of the intersection between feminism and biopolitics. Here the study will primarily be a discourse analysis of 'early' feminism, emphasising its interrelation with biopolitics and with a positive, transformative, political economy of experience. 'Early' feminist discourse will be read through the lens of recent historical scholarship on the issue of eugenicist feminism. My arguments depend upon my own interpretation of the discourse of eugenicist feminism, read in terms of the positivity of biopolitics – I am not simply deferring to (indeed I am often questioning) the historians' analyses. Nonetheless my reliance on secondary sources does mean that the picture of 'early' feminism that I am drawing is very much limited by the perspective of these particular recent historians and the choices that they have made in the selection of source material. This does not seem to me inappropriate because my ambition is to analyse a particular aspect of the discourse of 'early' feminism setting this in contrast to an alternative interpretation (that of Monique Wittig), it is not to establish or to question the historical event of eugenicist feminism and it is certainly not to develop a general appraisal of 'early' feminist thought. The historical studies upon which my analysis depends cover a wide range of areas of feminist history, from literature to the promotion of birth control, from Germany to the USA. They do, as such, grant access to a wide-ranging discourse, rather than to the particularity of text to which primary research would have led.

The second stage of the genealogy will interrogate the anti-biologism that was prevalent in second-wave materialist feminism, taking the limit case of Monique Wittig as an example and reflecting on her comments regarding 'early' feminism. Wittig opposed biologistic thinking about the character or specificity of women in general and presented her constructivism as a progression from 'early' and other feminisms that took up tenants of biologism. I will attempt to illustrate the problematic nature of Wittig's presentation of biologism as *conservative*, invested in being and stasis and extend my argument that biologistic discourse is about the allures of becoming not of being. Wittig is enabled and encouraged to engage in this presentation of the biologism of 'early' feminist thought as conservative because she adopts a *negative strategy of critique:* setting the difference of the past up as her *opposite* (defining her own position negatively, I=not-that); and attempting to prove the *fallacy* of the 'opposing' position. In this strategy, defining elements of biopolitical rationality fall into the gap that results from positioning biologism and the affirmation of becoming on opposing sides – defining elements that include the *progressive* dynamic, inclusive racism that is, according to Foucault, specific to biopolitical modernity. I will also attempt to capture something of the positivity of Wittig's own position – her own expansive, empowering affirmation of becoming. I will suggest that there are considerably greater parallels between her own position and that of 'early' biologistic feminism than her critical rhetoric would suggest, at least in terms of the economy of experiential allure and politics of temporality. I will further suggest that the negating critique of biological politics obscures the nature and real dangers of that politics, making it if anything easier for those dangers to be repeated in post-biologistic feminisms.

Before commencing with the genealogy and the 'chapter proper' I will outline the issues in contemporary feminist theory to which this genealogy pertains – issues that relate to the changing relationship between feminism and biology that is emerging in contemporary theory.

Biology, corporeality, racism and contemporary feminism

The relationship between social constructivism and biology is changing. We have seen, in Chapter 4, that Nikolas Rose is contesting the assumptions of sociological anti-biologism and demonstrating that, in post-molecular times, biology is a matter of artifice, hope and risk management (not biological destiny, discipline or normalisation). The change of perspective has been profound in the context of feminist theory. Whilst many feminists continue to denounce biological thinking as a patriarchal enemy (e.g. Jackson & Rees, 2007), an opposition to biology can no longer be assumed in the circuits of feminist thought. A number of post-structuralist feminists have rejected the sex/gender distinction and argued against the assumed normative and ontological priority of the 'cultural' 'discursive' or 'linguistic'. Many have argued that, whilst the anti-biology constructivist ontology of second wave feminism had much strategic value, it is far from having succeeded in the objective of eliminating the problems associated with essentialism. Whilst insisting upon the contingent and transforming nature of the cultural body – the body 'after' culture and socialisation – the distinction between sex and gender can be seen to compound essentialism and determinism about a body that exists 'before' culture. In being before socialisation and culture, the sexed body, in the sex/gender schema, is assumed to be before and beyond transformation, problematically affirming major tenets of essentialist determinism (Fuss, 1989; Haraway, 1991; Fraser, 2002:607–11) and reinforcing rationalist conceptions of the subject (Gatens, 1996:3–18) rather than emphasising the changing and open nature of corporeality itself (Grosz, 1994; Fraser, 2002), or the artificiality and abstraction of present knowledges and technologies of sex (Parisi, 2004). Within present post-structuralist feminist theory there is an effort to revalorise the corporeal and biological, so much maligned in the discourses of anti-biologism and social constructivism. This effort is coupled with a more or less vitalist (certainly *creative*) conception and celebration of the corporeal. Bodies, according to vitalist and corporeal feminists, are self-differentiating, open and creative.

Corporeal & vitalist feminism

Elizabeth Grosz has been one of the key figures in this reinvention of the body in feminist thought. Her most influential book *Volatile Bodies*, is a sustained address to the problem of unravelling the body/mind dichotomy and an exploration of the agency and indeterminacy of bodies. Like performative feminism and queer theory Grosz sees the work of feminism in terms of an opening up of indeterminacy, rather than as a politics of identity or a normative project (Grosz, 2004:260). Whereas performative feminism has (in Grosz's controversial view) situated possibilities for openness and creativity within the limits of culture, Grosz has concentrated on the analysis of the creative and historical agency of bodies in terms that do not refer the body back to a cultural or linguistic domain. Grosz emphasises the point that the rejection of the sex/gender distinction is not about saying that there is no sex and thus that all is temporal and contingent *because there is only culture* but, rather, illuminating the contingent and temporal character of the body and sex *themselves*. Sexed bodies are not beings, they are not even polymorphous beings, but 'polyactive becomings.' (1995:99). Biological organisation 'opens up and enables cultural, political, economic and artistic variation' (2004:1). In *The Nick of Time* Grosz engages with Charles Darwin's, Fredrick Nietzsche's and Henri Bergson's conceptions of evolution and philosophies of time. Here she reclaims Darwin as something of a difference feminist and develops a specifically *corporeal*-feminist interpretation of Bergson's conception of creative evolution. This conception elevates the powers of indetermination of bodies to the status of the forward momentum of time itself (2004:255).

Grosz does not only identify the corporeal as a domain becoming. In a move that explicitly follows Bergson's socio-biology she makes the becoming of corporeality into the normative *basis* of ethics and feminist politics. She adopts Gilles Deleuze and Felix Guattari's articulation of affirmative politics in terms of 'becoming'; becoming-woman, becoming-minority, becoming-imperceptible (Grosz, 2002). In this conception, becoming is differentiation, is active force, is expression. The differentiation of becoming is not the proliferation of categories so much as the deteritorialisation, the melting into air, of categories, the excessive production of expression and the generation and dispersal of power or active force. In an unacknowledged divergence from Deleuze and Guattari, however, Grosz ties the possibility of becoming specifically to corporeality; to the force of biological bodies. The body as such becomes the passage of time, the existence of difference and creative force. A politics of openness and difference – of becoming – is, for Grosz, a politics of the corporeal.

I have attempted, in this book, to foster a certain discomfort about such a politics – a discomfort akin to that we should feel about Rose's optimistic invocations of a politics of life itself. Whilst I accept Grosz's account of the corporeal and biological as domains of creativity and becoming (and have found her analysis of Darwinian evolutionary theory a great help in the background thinking of this book) I see her assumption of the desirability and political self-sufficiency of becoming as immensely problematic. Grosz's politics seems to be grounded in an infatuation with the creativity and diversification of evolutionary life and corporeal power. There are very significant differences between Grosz's imagining of corporeal politics and those of eugenicist feminists, but the positions do share in a conception of biology in terms of becoming and an infatuation with the creative force of evolution. It is not my intention to accuse Grosz of eugenicist racism (or of any other kind of racism), but I do want to insist that an affirmation of contingency, creativity and indetermination are no kind of safe guard against racist thinking. In contrast to Grosz I am emphasising the processuality and becoming of the biological, not in order to celebrate corporeality as creativity but to alert us to (and loosen the affective hold of) the biopolitical economy of experience as becoming.

In her adulation of the creative force of biological evolution Grosz says nothing of the human historical conditions of its production. Creativity, as such, (as in Bergson's socio-biology [see Hallward, 2006; Blencowe, 2008]) appears as a force coming into human life as though from a metaphysical-biological creative outside. For all her celebration of diversity and differentiation, life itself appears as a singularity in Grosz's writings. Grosz invites us to revel in the diversity and beauty of creative life, but she says nothing of the virtual multiplicity or technological producibility of life, nor of strategies for creating different formations of limit-experience, relationality or trans. There is a certain impotency about Grosz's politico-ethical vision. Picking up on Walter Benjamin's critique of Bergson (Benjamin, 1999b), we can suggest that the creativity of corporeality about which Grosz is so enthused may be little more than a romantic imagining of the passing moment (*Erlebnis*) as though it were the virtual creation of planes of limit-experience (*Erfahrung*); the difficult exciting drama of living together in the world (political action).

Rather than a turn to the body or to life, the positive critique of bio-politics points towards work upon the limits, the structures of experience and the productions of 'trans' from which multiple, historical, lifes might emerge. It is not a question of freeing life or embodiment from power or the stasis of forms, but of developing pragmatic grasps upon the structures, technologies, embodiments and imaginaries through which we are made as bodies desiring and becoming such freedom, transformation and affective capacity.

This approach resonates with the work of more 'Foucauldian' and 'Spinozist' vitalist feminisms, such as Moira Gatens and Luciana Parisi. Parisi treats creative life as multiply produced within human (and non-human) technological history. She criticises the singularity of Grosz's conception of life – pointing to the non-binary sexes of microbial and nanotechnological lifes (Parisi, 2009:46–7). Evolutionary life is multiple and technological. Parisi and Gatens both understand embodiment in *considerably* broader terms than that of Grosz's corporeality. For them 'embodiment' and 'experience' refer to the entire affective context (Gatens, 1996:131; Parisi, 2009: 48). Here the divisions that Grosz's arguments draw upon, between bodily creativity and linguistic form, are dissolved as embodiment is understood as imaginary and the imaginary is understood as material force. Rather than a Bergsonist opposition between creative-force and determined-form these thinkers invoke a flat ontology, wherein creativity is immanent to material determination. Freedom, in such imagining, is about the comprehension of, not the escape from, the determination of desire (e.g. Gatens, 1996:128).

Racism and contemporary feminism

As I will argue below, an opposition to, and the fear of complicity with, eugenicist and imperialist biological racism has been a significant concern for post-Second World War feminism. Such concerns, I will suggest (following Vikki Bell [1999]) are, in part, responsible for the anti-biologism that was dominant in the ontological wranglings of feminists for a number of decades. In the past decade the position of anti-biologism in feminist theory has, as we have seen, become far less secure. I want to suggest that the recent easing of tensions between feminist theory and biological sciences reflects a shift in the locus of such concerns and tensions from biological to cultural formations of racism; from biological to cultural biopolitics.

Racism (and class supremacism) persist as problems and fears for contemporary feminist philosophy and politics. In recent years the ethical impulse and vocabulary of liberal feminism has been taken up (and brutalised) in the voice of neoconservative militaristic imperialism, most notably in the discourse of the Bush Administration concerning its war in Afghanistan (Abu-Lughod, 2002; Power, 2009:11–17; Butler 2009). Whilst we might like to dismiss such self-evidently anti-feminist articulations as irrelevant ideological aberrations, debates such as that over the wearing of Islamic dress in France (where a strong feminist voice has supported the State in its attempts to exclude Islamic dress from public spaces) suggests that fears around and oppositions to Islamic culture and politics are immanent to strands of feminist thought. Perhaps the most intractable and ambiguous co-implication of feminism and supremacist or racist politics pertains to the issues of international development, wherein the 'emancipation of women', 'gender development' and 'reproductive rights' mark key sites of the potential slippage between the promotion of human rights or equality on the one hand, and the exercise of cultural domination and the perpetuation of global hierarchy on the other (Mohanty, 1988; Petchesky, 2003:31–64; Duffield, 2007:105–9).

I do not intend to suggest that all, or most, contemporary feminism is racist. Contemporary feminists are the sharpest critics of utilisations of feminist arguments to imperialist and racist ends. My point is simply that racism persists as a *problem* for feminism. Contemporary feminists, like their predecessors, worry about racism and about the potential complicity between feminism and racism, and sometimes feminist thought is racist or complicit with racist discourse. The biggest problem for contemporary feminism is a racism that is centred upon *culture* (rather than biology), especially upon Islamic and 'traditional' culture. A reimagining of the past relationship between feminism and biology harbours lessons for the way in which we deal with the problem of cultural racism in the present. In particular this genealogy attempts to break the circuit of assumption and affect by which the problem of racism and essentialism is associated with reification and determinism.

I would support post-colonial feminist Uma Narayan's claim that cultural relativism is no route out of Western Imperialism (2000:94–5). Narayan is right to argue that cultural imperialism can work as much through the production and presentation of difference as it can through the global imposition of the Same. She is certainly also right that the effort to promote global human rights is not an act of Western intellectual tyranny (Ibid.:91–3). However, I want to derail the assumption (that it seems Narayan makes) that once the problem of imperialist discourse has been identified as the assertion of cultural difference, then the battleground must centre upon the contingency and transformability of such difference. The problem with feminist and imperialist discourses of difference is, she suggests, that they draw 'ahistorical essentialist pictures of culture [that] obscure the degree to which what is seen as constitutive of a particular "culture" and as central to projects of "cultural preservation" *changes over time*' (Ibid.:88 – original italics). The solution, she suggests, is to insist upon and demonstrate the historicity and transformability of culture. Doubtless Narayan's call to constructivist arms has a basis in realities of reifying practices. I am not trying to suggest that cultural (or biological) reification is not, and has not been, a political problem. But I do want to insist that we should be open to understanding formulations of racism that deploy very different economies of experience: including biopolitical racism with its affirmation of process and change.

Insisting that cultures can change over time does not defuse a racist fragmentation of life that has been articulated in the context of technologies of liberal governance, immaterial production and creative transformation. Whilst Narayan rightly insists that Imperialism is not only about the imposition of the Same, the only alternative that she identifies is the production of 'the Other'. Either option – imperialist discourse as the imposition of the Same or imperialist discourse as the production of Others – seems to assume that the politics of race discourse has to be explained in terms of the logistics of identity; the universalisation of Western identity, or the production of Western identity through the construction of reified Eastern Others (Ibid.:83). But the notion of identity (and its implicit assertion of the analytical primacy of the need for epistemic security) fails to grasp so much of the experiential economics of subjectification. Classifications are technological, material things that make action possible. The allures of transformation, affect and influence also have a role to play in the explanation (not only the contestation) of racist constructs. When a Canadian development worker invests in a supremacist discourse concerning the difference between her own and 'developing' cultures she is not necessarily or simply producing an image of herself (against an Other) for the sake of a sense of security or self-satisfaction. She is also engaged in her self production as an autonomous moral subject, able and authorised to act (see Heron, 2007:99–104). At the least such authority cannot be dismissed as simple self-aggrandisement in situations wherein the imperative to act is an absolute ethical demand – the kinds of situations that are often faced by development and relief workers. A critical comprehension of cultural racism does not simply mean denouncing dusty old cultural essentialism or learning to live with transforming identities. It means situated pragmatic analytics of power/knowledge as it is enacted in actual relations; an analytics that is honest enough to recognise the situated *appeal,* affective force and productivity of supremacist discourse.

The genealogy that follows is both an attempt to grasp elements of that appeal in the context of feminist history and to loosen the affective force of present culturalist and vitalist affirmations of becoming, affirmations that could contribute to the stickiness of present supremacisms, by calling into question one of the dominant stories concerning the becoming of contemporary sociological feminism. The genealogy develops from Foucault's genealogical strategy to set out a positive-critique of biopolitics – contrasting this with the negating 'ideology-critique' adopted by Wittig and others.

'No Votes for Women – No Census': the biopolitics of 'early' feminism

Whilst feminists have mounted some of the most powerful challenges to biologisms, old and new, and rightly identified what we can now term 'biopolitical discourses' with forms of 'public patriarchy' (see Shu, 2006), feminism can itself, in significant parts at least, be described as biopolitical. This is perhaps most self-evident with respect to the 'early' feminism of the start of the twentieth century, wherein feminism, evolutionary thinking, biologism and eugenics were frequently intertwined. Illuminating the biopolitical character of 'early' feminism adds considerable substance Foucault's picture of biopolitics as the production of positivity, expansion of embodiment and experience of augmenting capacities.

An incident that has recently gained attention, with the release of the 1911 UK census data, is suggestive as a symbol of the connection between biopolitics and feminism at the start of the twentieth century. The modern census might be considered the *primary* biopolitical tool, making population-life apparent, observable, virtually governable. This is particularly so with the 1911 UK census, which was the first to ask questions concerning fertility in marriage, as the government sought to address the 'population problem' – which is to say, the need for a large and healthy population and fears over declining birth rates (1911-census.co.uk).

In 1911 the suffragettes of the Women's Freedom League mounted a boycott of the census. An estimated 'several thousand' women took part, either evading the census by staying away from home on the night of the count, or returning an uncompleted or spoilt form, often with posters attached or notes in the margins (1911census.co.uk). Comments and posters declared, 'If I am intelligent enough to fill in this paper, I am intelligent enough to put a cross on a voting paper', 'No persons here, only women', and 'No Votes for Women – No Census' (Meikle, 2009).

What this last slogan seems to capture is the sense that the rolling out of the biopolitical state – with its objectives of securing, maximising and regularising the biological life of its population – would have to pass not only through women's bodies, but also, to an extent, through women's *empowerment*. Whatever the actual historical consequences for women and gender equality (which are surely ambiguous and multiple) there was considerable virtual common cause in the strivings for women's empowerment and for the deepening and development of biopolitical governmentality. Given the increasing concern of the state with matters of family health and reproduction, some groups of women had an unprecedented access to public power and authority. If the state needs to know and control the intimacies of women's bodies and practices, a collective of women can strike directly at the needs of the state. Moreover, if reproduction and family health are considered matters of high politics – matters that will determine the very survival of an imperial Empire or the evolution a race – then the capacities, power and authority of women, as mothers and guardians of family health, can claim a radically augmented, and *public*, stature. The practices of mothering take on national, racial, truly historic, significance in the biopolitical gaze. In the boycott of the 1911 census we witness women making claim to that stature – to public citizenship – in the most biopolitical of potential sites of protest, literally refusing the state access to the population life that it is seeking to govern by not completing the census data.

Many feminists at the start of the twentieth century actively campaigned for biopolitical reforms – for, in effect, the deepening and intensification of biopolitical governmentality. In particular, the ideas and arguments of many feminists can be seen as prefiguring and informing the emergent welfare state. Historian Ann Taylor Allen argues that amongst early twentieth- century feminists in Germany, as elsewhere, there was an effort to open up the affairs of the family and private life to public scrutiny and public control; 'many women faced with the alternative between dependence on, and control by, the family or the state often chose the state, because it was more open to the collective pressure of the women's movement' (Taylor Allen, 1993:31). Amongst the examples she cites are the Abolitionists (the abolition of prostitution, that is), who in order '[t]o replace the traditional, patriarchal authority of the police, ... ([and] like other feminist reformers of their time) called for a new, motherly form of state authority – nurturing, protective, and controlling' (Ibid.:36). Feminists of the period (not unlike ecological feminists today) identified the ills of the state with its domination by men and foresaw a combined movement whereby women would be empowered and the state would be transformed into a caring and protective agency, acting to extend women's power and equality. In France Hubertine Auclert called upon women to transform the 'Minotaur' state, devoted to war, to the 'Mother' state, committed to the nurture of all its citizens (Ibid.:49).

Doubtless there was much naïveté in some 'early' feminists' support for the emergent welfare state as a feminist enterprise, as Taylor Allen suggests (1993:48–50). But there can be no doubt that key biopolitical developments constituted a radical empowerment of women with respect to their own bodies and life practices – the development and state distribution of the contraceptive pill in most Western countries being one obvious example. Moreover, biopolitical reasoning rendered public, powerful and historically significant matters of previously private, domestic concern. This created, or complied with, a new public, political and professional platform for women. As Linda Gordon has noted, state welfare has been overwhelmingly provided and received by women (1990:1).

Without this empowerment, without this making public-historic of the maternal role, body and domain, there could not have been the qualitative expansion – the deepening and refining – of biopolitical governance to which the twentieth century did in fact bear witness ... no public citizenship, no votes for women, no census, no biopolitical welfare state. And quite possibly the converse is also true – had there not been a biopolitical state interested in such things as fertility in marriage and public hygiene, there may have been no foothold from which twentieth-century Western women would establish their modern claims to citizenship. Whatever the plausibility of such speculation, there is without doubt a certain intimacy between twentieth-century biopolitics and the feminist movement. This close relationship between feminism and biopolitics is a historical testament to the positivity of biopolitics, including its association with radicalism and empowerment – as well as with a certain dynamic, 'evolutionary' racism.

The clearest document of the convergence between feminism and biopolitics is the incidence of feminist eugenics in the early twentieth century.

Feminist eugenics and the positivity of biopolitics

The relationship between feminism and eugenics that is by far the most familiar to contemporary social science is that of critique and contestation. One of the defining battles of the feminist movement in the latter half of the twentieth century, theoretically speaking at least, has been against biologism, which is widely regarded, in feminist circles, as synonymous with essentialism, determinism and sex discrimination. Eugenics, with its connotations of, at best, state control over women's bodies and, at worst, genocide, could well stand for all that is wrong in biologistic thinking from a feminist point of view. As a heavily gendered process, eugenics is the ultimate symbol of bad biopolitics, for women even more than it is for men. Indeed, feminist historians such as Jane Lewis have illuminated the patriarchal character of eugenicist politics in its early twentieth-century heyday and documented tensions between the proponents of that politics and their contemporary feminists (Taylor Allen, 1993; Lewis, 2000:86–88; see also Ziegler, 2008). The relationship between feminism and eugenics is, however, far more ambiguous than these factors would suggest. As a growing historical literature makes plain, feminism and eugenics were frequently intertwined, establishing the biopolitical character of early-twentieth-century feminism and demonstrating the positive, 'empowering' momentum of biopolitics. The political striving of these feminists for the improvement of the lives and stature of women was invested in broader, trans-organic politicised embodiments and life: those of the race, the nation and civilisation.

Eugenics

Eugenics – the improvement of the quality of the life of the human or racial species through the conscious and discriminatory control of breeding – was widely advocated in the UK and the United States throughout the last few decades of the nineteenth century and up until the Second World War – at which point eugenicist policies became associated with the Nazis and much more difficult to support, explicitly at least (Taylor Allen, 2000:499; Ziegler, 2008:233–4). By 1900 eugenicist ideas had spread from the UK and the US to Sweden, Norway, Russia, Switzerland, Germany, Poland, France and Italy (Lewis, 2000:82; Mottier & Gerodetti, 2007). Eugenics was a radical and 'progressive' politics which drew heavily (if somewhat selectively) on Charles Darwin's evolutionary biology, especially upon: his theory of *natural selection*; his powerful biological conception of *evolution;* and his statement of the reality and evolutionary functionality of *variation within species.* Eugenics was founded by Francis Galton, who saw human variability as the potential source of racial progress and advocated the self-conscious control of human evolution through selective breeding (Richardson, 2000:44). Eugenics was a matter of transformation and becoming, not of conserving the status quo – it pertains to limit-experience, pushing at the limits of present possibility, not (as is supposed by Wittig) to conservative experience that draws upon the strength of tradition.

All advocates of eugenics favoured 'eugenically responsible' control of reproduction, preventing or discouraging the unhealthy from having children so as to enhance the future – more evolved – life of the species. 'Health' was writ extremely large in this discourse such that the disabled, the mad, the criminal, the poor and even ethnic minorities could figure as unhealthy or 'degenerate' – to be discouraged from breeding or at least from interbreeding with and polluting 'the fit' (Weikart, 2004:71–126). Galtonian eugenics was particularly interested in class (conceived in what we would now think of as racial – that is, biological – terms) and 'aimed to regulate the population by altering the balance of class in society' (Richardson, 2000:44).

Theories and policies concerned with reproduction intertwined with various discourses concerning social hygiene. In such discourses a conflation was frequently made between physical health and the morality of conduct. Both illness and immorality were understood on the model of contagious disease and, crucially, as contagious across generations. Immoral behaviour could result in illness or deformity in future generations. Philanthropists and reformers sought to deal with disease in cities through programmes of moral rectification and education; especially of urban working class mothers. Such campaigns were accompanied by the first rungs in the establishment of state welfare, such as financial support for mothers (Lewis, 2000:86).

The early advocate Ernst Haekel suggested that eugenics should extend to the infanticide of 'miserable and infirm children' whose preservation, he maintained, would be to 'their own harm and the detriment of the whole community', thus prefiguring the monstrous radical applications of negative eugenics in the Germany of the 1930s and 1940s (Haekel, 1870, cited in Weikart, 2004:146). Whilst most nineteenth-century eugenicists would certainly not have endorsed such arguments, the very logic of eugenic reasoning does radically subordinate the life of existing individuals to that of the future collective, 'the species', and differentially distributes existing classes of the living in relation to that future life. As such, the radical subordination of the lives of some existing groups to that of others is legitimised, indeed, *required* by eugenicist politics – however moderate, compassionate or left wing (and eugenics was at least as big on the left as on the right [Franks, 2005:26; Mottier

& Gerodetti, 2007]). Whether the salient differentiation is that of race, class, health or morality eugenics is necessarily a supremacist politics.

Feminist eugenics

Eugenicist ideas were immensely popular amongst turn of the century western feminists, becoming increasingly so through the first decades of the twentieth century. These early feminist movements were not only centred on the individual rights of women 'but also on the purification and regeneration of society' which was identified by many feminists as women's most important mission (Taylor Allen, 2000:477; 1993:27). Campaigns for women's rights to birth control, humane working conditions, maternity leave and welfare provision were at times supported by both feminist and non-feminist eugenicists and publicly justified through eugenicist arguments (Weikart, 2004:133–4; Lewis, 2000:83; Ziegler, 2008). At the same time a medicalised morality, suffused with scientific ideas, rhetoric and authority, offered educated feminists an exciting and progressive alternative to traditional, patriarchal, codes. As Taylor Allen puts it, 'eugenics offered [feminists] the exciting possibility of replacing the traditional Christian morality, based on sexual repression, with a new morality based on a rational understanding of natural laws' (1993:31).

In the UK, as Angelique Richardson documents, the 'New Women' that wrote and populated more than a hundred English novels in the last two decades of the nineteenth century, were supporters of Galton and advocates of eugenics (Richardson, 2000:45). Indeed, for some of these women feminism could be *equated* with eugenics. Sarah Grand, one of the best selling of the New Women authors,[3] cites taking good charge of matters of population, the improvement of the race and thus 'saving our present civilisation from the extinction which has overtaken the civilization of all previous peoples' as the very *purpose* of women's empowerment and suffrage (Grand, cited in Richardson, 2000:45). As Richardson puts it, '[e]ugenic feminism might coexist with the desire to emancipate women from patriarchal law, but it sought to replace that law with an authoritarian health regime' (2000:46).

Such feminists were not blind to the implication of negative eugenics. Grand has one of her heroines, Beth, declare medical help for the 'unfit' an unwelcome hindrance to Nature's good evolutionary work:

> Nature decrees the survival of the fittest; you exercise your skill to preserve the unfittest, and stop there – at the beginning of your responsibilities, as it seems to me. Let the unfit who are with us live,

and save them from suffering where you can by all means; but take pains to prevent the appearance of any more of them. By the reproduction of the unfit, the strength, the beauty, the morality of the race is undermined, and with them its best chances of happiness. (*The Beth Book* 1897, cited in Richardson, 2000:46)

In the first years of the twentieth century in Germany, according to both Richard Weikart and Ann Taylor Allen, many leading feminists, such as Ruth Bré and Helene Stöcker were 'staunch supporters of eugenics' (Taylor Allen, 2000; 1993; Weikart, 2004:97). Bré and Stöcker were founding and leading members of the League for the Protection of Mothers, established in 1905 and concerned, in the first instance, with the welfare of illegitimate children. Their campaign was fought with reference, not only to individual welfare, but also to the health of body politic at large. In the group's first manifesto they 'castigated a society that expended huge resources on the "sick and Sterile" but allowed many of its most "valuable offspring" to die' (Taylor Allen, 1993:34). Their assistance would be offered exclusively to 'healthy mothers' (Weikart, 2004:133).

Again, the eugenicist goal of improving the quality of the race is a considerable part of the *justification for* the emancipation of women – in particular for the empowerment of women with respect to birth control. Stöcker was one of the earliest leaders of the birth control movement. Whilst sexual liberation was important for her, she also hoped that widespread use of birth control would enable women to follow their 'moral duty' not to reproduce if the interests of the future generation would be harmed (Weikart, 2004:133). Moreover, one of the prime benefits of widening access to birth control would be to slow down the reproduction of the 'inferior' lower classes whose offspring threatened to swamp society with their bad heredity (Ibid.:134). And again, these feminists were no stranger to the cause of negative, eliminative, eugenics. Stöcker declared 'You know that there are many cases in which a woman is not only justified, but positively obliged to terminate her pregnancy; in case of contagious disease, mental illness or alcoholism' (cited in Taylor Allen 1993:40–41).

Such concerns were not limited to European feminists. Charlotte Perkins Gilman, widely regarded as the leading intellectual of the woman's movement in the US at the turn of the century, was quite explicit that the purpose of sex is not recreation, 'free and selfish indulgence', but 'the conscious improvement of the species' (Gilman, cited in Scharnhorst, 2000:70). Likewise, the leading proponent of birth control in the United States, Margaret Sanger, stated in 1919: '[m]ore children from the fit; less from the unfit – that is the chief issue of birth control' (Sanger, cited in Weikart, 2004:135; see also Franks, 2005). In 1918 she wrote:

> All our problems are the result of overbreeding among the working-class, and if morality is to mean anything at all to us, we must regard all changes which tend towards the uplift and survival of the human race as moral. Knowledge of birth control is essentially moral. Its general, though prudent, practice must lead to a higher individuality and ultimately to a cleaner race. (Sanger cited in Ziegler, 2008:229)

As in Europe, the distribution of benefits ensuing from eugenic policies was discriminatory. As Lewis notes, 'efforts at enhancing childcare were largely directed at white urban immigrant populations in the early 1900s and later at rural white women, to the neglect of African American women' (2000:86).

Indeed, the relationship between US feminist eugenics and purportedly 'unfit' communities did, in some cases, extend beyond a matter of mere 'neglect' to support for active persecution. Gilman, for example, was heavily invested in supremacist ideas concerning racial evolution and abhorred mixed raced partnerships. Among the proposals of the early-twentieth-century eugenicist movement that she endorsed was compulsory sterilisation for 'the unfit', whom she once defined as 'admitted defectives living on our taxes' (Scharnhorst, 2000:68–9). In 1915 she wrote:

> How about idiots? They are no good to themselves or to anyone else, they are, on the contrary, an injury.... We talk of 'the sanctity of human life[,]' and we are right. Human life is sacred, far too sacred to be allowed to fall into hideous degeneracy. If we had proper regard for human life we should take instant measures to check the supply of feeble-minded and defective persons. (Gilman cited in Ziegler, 2008:227)

Significantly, feminists who were involved with eugenicist politics did not simply fall in with the dominant eugenic science of their time. They redefined eugenic science, carving out a distinctive position of *feminist eugenics* that cast men's domination over women as a dysgenic (counter-evolutionary) force. This gives considerable support to Foucault's view of biopolitics as engaged in transformative limit-experience, because it suggests that these feminists were not conforming to the existing tradition when they adopted biopolitical eugenics, but rather creatively defining their own radical politics. Eugenics was a part of the radicalism and feminism of feminist eugenics, not a contradictory conservative force.

As Mary Ziegler maintains, whilst there were a variety of visions of feminist eugenics articulated between 1890 and 1930, 'each found a commonality in the argument that the eugenic decline of the race could be prevented only if women were granted greater political, social, sexual and economic equality' (2008:213). In her 1920 *Woman and the New Race,* for example, Sanger argued that it was women's socially and legally enforced ignorance about sex, sexuality and birth control that had given rise to the racial decline observed by eugenic scientists (Ziegler, 2008:230). Similarly, in Germany, Anita Augsprurg claimed that civilisation was poised over a 'stinking swamp of corruption', represented in the prevalence of the 'racial poison' venereal disease, which served as a metaphor for men's oppression of women; the cure for the corruption, and thus for venereal disease, would be the restoration of women to their rightful role in culture (cited in Taylor Allen, 1993:30). In the UK 'New Women' conflated eugenic policies with love, feminising eugenics (Richardson, 2000:46).

The ambiguous relationship between feminism and eugenics

The complicity between much 'early feminism' and eugenics does not detract from the larger fact that mainstream eugenicist thinking and policy was strongly patriarchal and anti-feminist. Leading eugenicist thinkers were invariably sexist in their beliefs, be that in the misogynistic tones of Otto Weininger, who declared that whilst 'a man possesses sexual organs, her sexual organs possess woman', or the more valorising vision of Patrick Geddes and J. Arthur Thompson who lauded women as 'eupsychic inspirer and eugenic mother' (cited in Lewis, 2000:86–87).

Such patriarchal beliefs fed a mainstream eugenicist position whereby it was argued that women, or rather 'fit' women, should reserve their energies exclusively for the work of motherhood. Support for that position blocked and questioned middle class women's access to higher education and working-class women's access to the labour market. Feminist eugenicists objected to such 'sexual sciences' (Lewis, 2000:86), and redeployed eugenicist reasoning against them. For example, between 1910 and 1916, Gilman published articles providing a eugenicist critique of contemporary norms of sex inequality that were actively supported by mainstream eugenicists in the United States. In particular she attacked social prohibitions upon women attaining a full education. These prohibitions were, she argued, directly dysgenic (counter-evolutionary); they crippled, stunted and atrophied the brains of women and thus their offspring. They were also indirectly dysgenic because more educated women would have fewer children, which would lead to higher quality offspring to the benefit of the race (Ziegler, 2008:226, 228). Moreover, negative eugenicist practices of sterilisation were very strongly gendered. The 'unfit' that were the target of eugenicist ideas and laws did not only include people of an inferior racial stock (indicated through phenotype, class and health) but also women who were perceived to be deviant, 'licentious' or neurotic (Ibid.:214). This is reflected in the more or less forced eugenicist sterilisation campaigns of countries such as Sweden, wherein almost 95 percent of eugenicist sterilisations were carried out on women (Mottier & Gerodetti, 2007:43). Again, many feminist eugenicists departed from mainstream eugenic theory, challenging such patently anti-feminist assumptions and policies – policies that included women in the category of 'defective' on the basis of their sexual behaviour, their lack of femininity, or, in the case of mental illness, relatively far less severe symptoms than was the case for men (Ziegler, 2008:233). The fact that there were feminist eugenicists certainly does not mean that eugenics was normally feminist.

The complicity between feminism and eugenics does, however, illuminate the ambiguity of the relationship between feminism and biologistic discourse. Women and feminists had a range of complicated stakes in the eugenicist movement, despite its frequently patriarchal implications and promotion of sexist beliefs. The support of many early feminists for eugenic ideas and policies cannot be simply put down to conformity to mainstream or traditional ideas of the time, not least because these feminists did define their own distinctive brand(s) of eugenic theory (Ziegler, 2008), and, whilst eugenic thinking might have been immensely influential, it was certainly *not* traditionalist (Lewis, 2000). There was a radical dimension to eugenicist thinking, one that challenged Victorian models of patriarchy (even as it sought to install new, more public, models) and described routes to empowerment for women – for middle class, white, 'healthy' women at the least. That radical dimension best explains the attraction of women to eugenicist ideas – including the attraction of enfranchised modern women to the eugenicist ideas of parties such as the British Union of Fascists in the 1930s (Shu, 2006:266). Crucially, we should note the apparent coexistence of that radical dimension, that progressivism, with a strong racial supremacism that encompassed hierarchies of class, health and moral virtue as well as phenotype. That supremacism was only consistently challenged by eugenic feminists in so far as it pertained to specific moral criteria that cast women as inferior and subject to men. The feminist challenge to patriarchy, in these instances, did not incorporate a challenge to biopolitical racism.

We can see, then, that 'early' feminism was engaged with emergent biopolitical politics, and not only with the 'caring', welfarist work of biopolitics. At least some feminists were invested in the eugenicist ambition of improving the life of the species, race, nation or civilisation through the exclusion or elimination of certain fragments of the population. Whereas later anti-biologistic feminists, such as Wittig, would argue that this association between feminism and biological theory was a result of a need for conservatism, a need for traditional authority or security, the history of eugenicist feminism make it plain that biological knowledge gave shape to the very radicalism and dynamism of the 'early' feminists' ontologies.

Eugenicist feminism and the positivity of biopolitics

The caricature of biological thinking and politics as conservative (which Foucault's positive critique of biopolitics, set out in this book, draws in to question) is played out in a number of accounts that are given of these 'early' feminists in feminist literature. In 1993 Ann Taylor Allen observed that these tendencies in early-twentieth-century feminism – in particular the campaigns against venereal diseases – expressed a 'bewildering and seemingly contradictory combination of emancipatory and repressive ideas' (1993:29). In 1980, Judith Walkowitz attempted to explain this apparent duplicity by suggesting that the original impulse of these feminists was emancipatory whilst the repressive aspects of their programme could be attributed to the influence of male dominated social purity movements (Walkowitz, 1980:246–57).[4] In a similar vein, materialist feminists in the 1970s and 80s – to be discussed at length below – attributed the 'contradictions' to the difficulties of radicalism and comforts of tradition or essentialism. Darwinism, Rosalind Rosenberg argued in 1975, provided early feminists with 'the reassurance they needed as an anchor of certainty in a time of social flux, an anchor that could protect them not only from critics' hostility, but just as importantly from their own uncertainty' (1975:142). Darwinist beliefs served, she seems to suggest, as a traditional, essentialist security blanket for these women who were not quite ready to embrace a fully emancipatory (which is to say constructivist) ontology.

Such interpretations, bent apparently upon safeguarding the emancipatory credentials of the 'early' feminists, are not only potentially patronising, they also obscure and contradict the historical record. In fact Darwinism, biologism and eugenics were associated with iconoclasm, radicalism, anti-traditionalism and non-conformity amongst the 'early' feminists, as in other groups (Taylor Allen, 1993:31, 48; Lewis, 2000). Moreover, there is nothing to suggest that these women were incapable of challenging either traditional or mainstream ideas, given that they did, in fact, energetically combat biased and misogynistic ideas from wherever they arose (including within eugenic science), whilst developing their own distinctive articulations and visions of eugenic science and practice (Taylor Allen, 1993:29; Zielger, 2008).

In line with the genealogical strategy of 'positive critique' we can, instead, interpret the pull of eugenicist ideas for feminists in terms of the production of force and embodiment that they engender – which is to say, in terms of their positivity. Rather than positing a desire for conservative experience (which seems to contradict what, in general, these women were about) we can identify the ways in which biologism created limit-experience, experience of going beyond the present limits of the self and the possible. We can begin to understand the coextensive appeal of feminist and eugenicist discourses in this period when we consider the relationship between these ideas and the expansion of forces, affect, embodiment, and thus of the experience of power. Whilst eugenicist discourses inflicted and supported much 'repressive' or counter-emancipatory influence for women (as for most men), they did also carve out expanded fields of action, embodiment and power, whilst promoting and participating in vitalist aesthetics of process, evolution and creativity. There was a considerable positivity in these discourses for feminists at the time, a considerable production of affect, embodiment, experience of power and potential action – a positivity that could resonate with the feminist drive for women's empowerment.

Expanding women's embodiment

In breaking down barriers between public and private bodies and fields of action, eugenic discourses choreographed new forms of embodiment in which the affective capacities of individual women were multiplied exponentially. Theories of reproduction, heredity, evolution, racial health and degeneracy transformed and multiplied the potential influence – the capacities – of bodies, especially in motherhood. All women, as apprehended in eugenicist theory, had a radically expanded historical and social significance as mothers and potential corruptors or redeemers of not only their own children but, through networks of heredity and infection, the life and health of entire cities, nations, races – even of the species itself or the Imperial Empire. Sarah Grand went so far as to declare that women's action in the cause of social hygiene was the 'only hope' of saving present civilisation from extinction (cited in Richardson, 2000:45). Within eugenicist discourse a traditional morality of good order gave way to an emphasis on the *public implications* of private choices (Taylor Allen, 1993:37). The private use of bodies became of profound public significance, which also meant, of course, that private matters became of increased public concern. Thus the eugenic recomposition of women's embodiment accompanied a series of new investments in women's bodies and practices – especially in their capacities as mothers – including welcome welfare provisions (such as the mothers' pensions legislated for in 40 US states by 1920) and less unambiguously welcome interference (such as that of the voluntary lady visitors in British cities, who were ridiculed by working class women in literature of their time) (Lewis, 2000:86).

For the elite women who were often at the forefront of feminist organisations, heading campaigns for birth control, state child care and the abolition of venereal disease, the range of capacities – the embodiment – presented by the role of 'mother and guardian of race, species or nation' could incorporate the bodies, practices and capacities of all those other women who were (potentially) touched by the fledgling welfare provisions, moral rectification and educational programmes that these feminists campaigned for or carried out. Work upon other women's mothering constituted a potential public role for middle class women of the time (Richardson, 2000:45). In breaking down barriers between public and private embodiment, describing individual bodies as participants in the life of population, eugenicist theories of heredity, degeneracy and social hygiene offered an expansion and intensification of embodied life – of affective capacity – to feminist women: women who were seeking, precisely, an expanded field of action and public citizenship.

Creating radical authority

Further to this, eugenicist ideas, and evolutionary theory more widely, engendered and augmented new forms of moral and epistemic authority – forms that could challenge tradition, Christianity and patriarchy, embracing a modernist aesthetic of transformation, process and vitality. Drawing upon eugenic reasoning enabled feminists to claim the authority not only of science, but of evolutionary life itself, as they fought to break the hold of traditional morality and patriarchal authority. Swedish writer Ellen Key, for example, defended women's access to careers through the argument that love matches would produce eugenically healthy children. Love, Key argued, is an evolutionary tool of natural selection. In forcing women into loveless marriages, traditional society flies in the face of the evolutionary drive of nature. Women, according to Key, should be enabled to become economically independent because then they would only marry for love and *therefore* act in line with the evolutionary forces of nature (Taylor Allen, 1993:37). Eugenic science did not so much provide an anchorage for radical women in an unchanging identity or traditional authority as assist women in establishing immanent and creative forms of authority, precisely as they *broke with* tradition and embraced an ethos of progressive change. Eugenic science and evolutionary theory were productive of virtual legitimacy for the creative, iconoclastic, transformative force of feminist – as other 'progressive' – endeavour and value formation.

Underscoring the positivity of eugenic reasoning for feminism is the conception and valorisation of life that is specific to modern evolutionary biology. Immanent, trans-organic, population life – not an eternal order or ancient past – acts as the outside anchorage, the external beyond, that reflects auratic power and authority upon eugenicist ideas and action. Life constitutes *experience*: processes of escaping finite singularity and productions of depth and connectivity in the present. It does so without valorising and manifesting an eternal order or sanctified past. If it is nature that serves as a legitimating authority for these feminist eugenicists it is not the 'vegetable nature' of classical thought – nature as a pre-given and total order of fixed relationships. It is, rather, the 'animal nature' of modern biology – a nature of 'eternal becoming' (Stöcker cited in Weikart, 2004:132) that is in perpetual transformation, proceeding through self-transcendence and self-destruction. An image of evolutionary life as a contingent, transforming, *creative* force acted as inspiration and support for radical feminists, expanding, not curtailing, their field and ferocity of virtual action.

To be clear, my argument is not that eugenic reasoning was 'good', nor that it significantly, let alone primarily, acted in the interests of women and their emancipation. It is rather that, for some privileged and politically radical women at the least, eugenic ideas reconstituted embodiment such that their bodies – their affective capacities – were expanded in space and time. Eugenic and evolutionary theories and practices reconfigured embodiment such that many women, especially elite and radical women, could experience an expansion of capacity. At the same time these ideas could augment, add weightiness or depth to, radical efforts at value *creation* – informing and endorsing an ethics, epistemology and aesthetics which was addressed to a world in indeterminate transformation; evolution. In recognising the genuine appeal of eugenic reasoning for 'early' feminism we are not endorsing that project. We might, however, move towards a clearer sense of the affective dynamics of both biopolitics and feminism in the early twentieth century, tuning in to notes of the resonance between them.

Feminist anti-biologism

Having established the positivity of biopolitics and eugenics in the specific instance of early-twentieth-century European and northern American feminist discourse we will now move forward to the 'post-molecular' context that was the subject of Chapter 4. In Chapter 4 we established that there was a widespread anti-biologist turn, especially against eugenics and race science, in the wake of the Second World War. This anti-Nazi anti-biologism took place across the political spectrum (although anti-racists and feminists positioned themselves against their contemporary right whilst neo-liberals positioned themselves against the spectre of Communism). Second-wave feminism was a major contributor to that anti-biologist turn. In the remainder of this chapter we will elaborate upon the problems with 'negative critique', as deployed by anti-biologist constructivist feminism, as a strategy for addressing the problems with biological politics. We will also suggest, in line with arguments of Chapter 4, that radically constructivist feminism might be seen as a reinscription rather than a rejection of (some aspects of) the biopolitics of 'early' feminism.

In the 1970s and 80s, the context of 'second-wave' feminism, the relationship between feminism and biological discourse was largely characterised as one of conflict and contradiction. A central theme in feminist theory of this period was the contestation of 'biological determinism' and the advocacy of social constructivism – an ontology that was largely understood to be emancipatory in and of itself. Biology, in feminist circles, was usually associated with essentialism, conservatism and an insistence upon the natural or unchangeable character of differences and inequality.

Establishing a distinction between sex – as given bodily difference – and gender – as socially constructed difference – was one of the major strategic moves of second wave feminism, enabling feminists to counter claims that given relationships and differences between women and men were 'natural' and thus legitimate or unchangeable. According to Donna Haraway all the feminist meanings of 'gender' have their roots in Simone de Beauvoir's statement that 'one is not born, but rather becomes, a woman' (Beauvoir, 1960: 8; Haraway, 1991:130; Fraser, 2002:607). Debates around the respective importance of biological and socio-cultural influences in the determination of sex difference – the debate between nature and nurture – were animated by this ethico-politically inflected temporal dualism of being and becoming: Nature as fixed essences; Nurture as becoming, transformation, construction. The discourse of the sex/gender distinction established the cultural body as a site of change and transformation, whilst eliding sex and biology with stasis, essentialism and determinism. As Mariam Fraser argues, this elision of sex with biology and biology with essentialism in fact *presupposes* that the biological body is politically and materially static and that the cultural body is politically and materially malleable. It also continues and extends the distinction between interior and exterior, locating the possibility for change in environmental influence (Fraser, 2002:610–11).

The critique of biologism – and of essentialism, with which the term became synonymous (Birke, 2000:2) – was not just any old critique. Reportedly genuine 'paranoia' and 'paralysing anxiety' surrounded the issues of biology and essentialism in second-wave feminist circles (Bell, 1999:114–9). Of essentialism Diana Fuss was able to assert, in 1989, that 'few other words in the vocabulary of contemporary critical theory are so persistently maligned, so little interrogated and so predictably summoned as a term of infallible critique' (1989:xi). Even those 'difference feminists' who maintained the reality of a universal difference between men and women were keen to distance themselves from biologism and biological determinism (e.g. Gatens, 1996:3–18). There was something deeply emotive, immensely significant, even 'foundational' about anti-biologism for much second-wave feminist theory (and even more so for that of deconstructionist feminism of the 1980s and 90s).

Racism and anti-biologism

Feminist critiques of biological and naturalizing ideas were undoubtedly focused upon the relationships between sex and gender, men and women, hetero- and homo-sexuality. However, feminist anti-biologism did not exist in a vacuum. A range of problems motivated, informed and found resonance with feminist thinking and writing on the 'evils' of biological and naturalizing knowledges. One of the most significant was the problematic of race science, eugenics and the association of biological reason with fascist and racist politics. Feminist ideas about the ideological force and problem of naturalizing discourse were informed by struggles with and against racism and fascism, with a common narrative of oppression frequently invoked.

As Vikki Bell argues, the feminist imagination that de Beauvoir bequeathed younger feminists was populated by a range of rhetorical figures, including that of the black male struggling against racist oppression in the United States (Bell, 1999:42). Bell notes the warm friendship between de Beauvoir and the black American writer Richard Wright, as well as the intertextual resonances in their work. In Wittig's writings we encounter the rhetorical figure of African slavery. In 'The Category of Sex' (1996) and 'One is Not Born a Woman' (1993), where she articulates her radical opposition to biologistic or naturalizing claims (implicitly), African slavery furnishes Wittig with analogies for her statements about the domination of women through naturalizing ideology, as well as its potential overthrow. 'The perenniality of the sexes and the perenniality of slaves and masters proceed from the same belief' (1996:24); while 'lesbians are escapees from our class in the same way as the runaway slaves were when escaping slavery and becoming free' (1993:108, also 105).

In critiquing biologism, materialist feminists can be seen to be seeking a safe distance from the sorts of biological rationality and reasoning that legitimated the practices and hierarchies of European eugenicist politics and imperialism. As Bell argues:

> [i]f the 'social' is always privileged over the 'natural' in constructionist accounts – even when the shape of argument may not be dramatically altered and may need, consequently, more thought – it is…in part because such a privileging enables a distance to be taken from those times when bodies were so dramatically essentialised, when people were reduced to an essence as part of an order that placed them within cruel and deadly hierarchies. (1999:116)

Of course the idea that racism was a problem and source of paranoia for feminist theorists does not mean that feminist discourse and practice has not, in fact, frequently *been* complicit with structures and practices of racism. Indeed we can imagine that it is precisely such actual and potential complicity that generated the kind of crippling anxiety and paranoia for feminists thinking (or not thinking) the biological. What I do want to suggest is that the feminist articulation of anti-biologism in the 1970s should be seen in a broader discursive political context, a context wherein biological conceptions of collective life were tainted with the stain of fascism. A seismic discursive event took place in the decades following the Second World War whereby race science and socio-biology were roundly denounced, sociological (as opposed to socio-biological) knowledges gained a new authority and constructivist ontology was lauded as a form of resistance to fascism, racism and totalitarianism. That event included the shift in authority on matters of race difference from the natural to social sciences, as well as the development and coming to dominance of neoliberal rationality (Foucault, 2008:201–2; Skinner, 2007). The paranoia, anxiety and hostility concerning biology in feminist discourse can be understood as a part of that event. At the least, the role of biological knowledge in the production and justification of European imperialism, negative eugenics and fascist racism must be seen as elements in the fuel and formation of feminist anti-biologism.

Arguably, however, the anti-biologism of second-wave feminism did not succeed in establishing a distance from the experiential economy of biologistic naturalising discourse that were operationalised in earl-twentieth-century European eugenics, fascism and racist reproductive and heterosexist policies.

Certainly the shadow of imperialism and racism continue to hang over contemporary feminism, all be it in a newly 'culture-centric' guise. Islamaphobia, Western imperialism and cultural domination are live problems for contemporary feminists – problems feminist wish to address but also problems in which feminist discourses have become embroiled (Mohanty, 1988; Abu-Lughod, 2002; Power, 2009; Butler, 2009). The problems of difference and of supremacism recur throughout waves and movements of twentieth and twenty first-century feminist theory.

I want to suggest that some part of the responsibility for that repetition lies in the failure of feminist critique to get to grips with specifically *biopolitical* racism – the racism that is compatible with modernist, progressivist and sometimes feminist thinking in a world of contingency. Arguably the science of nature that anti-biologist feminism took as its target was not in fact the biologism (rationalities of modern evolutionary biology) that informed eugenicist and imperialist politics, but rather an older mode of reasoning about nature – that of natural history, the classical era and the police state – in which nature is understood as a preformed, fixed and completed order, a mode of reasoning that was confounded not compounded by Darwin.

The discourses of modern biology that informed eugenicist and imperialist politics that are made visible by the positive-critique of biopolitics were, in fact, discourses of *becoming* not of *being*, associated (in their own time) with iconoclasm and political radicalism, fascinated with an image of the world in indeterminate transformation. As Foucault argues (see Chapter 1 above), this was resolutely *not* the naturalism of preformism transfixed with the inherent unchanging total order of the world – an ontology that could accurately be described as 'deterministic' and conservative (an ontology that dominated the natural history of the early modern, 'classical' era).

The second-wave feminist critique of biologism was a 'negative' critique, in the sense of being a critique that sets itself up in *opposition* to its target – emphasising differences rather than continuities between positions, creating a plethora of dualisms and often seeking to morally condemn before it sought to understand. The critique of biologism, and early-twentieth-century biopolitical discourse, included its moral condemnation and supposed falsification as well as the assertion of the value and truthfulness of social constructivism as the *opposite* position. I suggest that this negative strategy of critique did, in *some* respects, backfire, creating a series of masks and miscomprehensions behind which the intended target – biopolitical racism and supremacism – could *hide,* escaping the anti-eugenicist critique.

Monique Wittig's anti-biologism and critique of 'early' feminists

I will now move on to develop these suggestions through the reconsideration of one of the most concerted of the second-wave feminist critiques of biologism – that of Monique Wittig. Second-wave feminism was nothing if not multiple. No strand or thinker can begin to stand in for the movement. However, Wittig's critique of biologism makes for an appropriate event in which to rethink the second-wave feminist critique of biologism precisely because of its very exceptional – somewhat exaggerated – nature. Whilst certainly not archetypical, Wittig was perhaps 'ideal typical' when it came to the issue of biology – articulating an exaggerated version of the anti-biologism that pre-empted the developments of deconstructionism that were to follow in the late 1980s and 1990s. She defended an (at the time) extreme form of social constructivism, drawing upon Marxist philosophy to characterise genders as social *classes* and biology as a crucial component of patriarchal and capitalist ideology. It is, then, Wittig to whom we will refer as an ideal-typical example of the second-wave feminist critique of biologism. After setting out the general tenor of Wittig's anti-biologism I will examine her critique of the biologism of early feminism and explore the contention that key elements of biologism and biological racism are hidden rather than addressed in this critique.

Wittig's anti-biologism

Wittig was one of the leading exponents of materialist feminism in France in the 1970s and America from the 1980s onwards. Her article 'One Is Not Born a Woman' appeared in 1981 and is a major reference point in the recent history of feminist theory, not only as an influential article in its own right but also as a major departure point for Judith Butler's performative, genealogical feminism (Wittig, 1993; Butler, 2004; 1999). Alongside Christine Delphy and others, Wittig was a member of the materialist-feminist *QF* collective, which produced the journal *Questions Féministes (QF)* between 1977 and 1980 (Jackson, 1995).

For the *QF* collective, all naturalistic ideas and explanations of sexual difference derive solely from patriarchal reasoning and the oppressive social structures that it legitimates. Indeed, 'men' and 'women' are to be understood on the Marxist model of *classes* (Jackson, 1995:13). Opposition to all naturalistic explanations of sexual difference is, they declared, a basic tenet of politically radical feminism. Any non-social notion of 'woman' or 'sex difference' should be refused and deconstructed, for both 'are an integral part of naturalist ideology' (*QF* collective, cited in: Jackson, 1995:14–15). The social mode of being men and women is, they maintained, in no way linked to their nature as sexed bodies.

Wittig insists that both materialist-feminist analysis as theory, and lesbian society as practice, demonstrate that the apparently natural group 'women' is an ideological construction – a product, not cause, of oppressive and unequal power relations. Lesbian society 'pragmatically reveals that the division from men of which women have been the object is a political one and shows that we have been ideologically rebuilt into a "natural group"' (Wittig, 1993:104). Wittig continues:

> In the case of women ideology goes far since our bodies as well as our minds are the product of this manipulation. We have been compelled in our bodies and in our minds to correspond feature by feature, with the *idea* of nature that has been established for us. (Ibid.)

In opposing naturalistic conceptions of differences between women and men Wittig understands herself to be opposing not only patriarchal ideology as espoused by patriarchs, but also as espoused by fellow feminists who have, themselves, fallen into the trap of the 'myth of woman'. The idea of natural differences and determination is patriarchal ideology and a conservative force of oppression, whoever gives it voice and for whatever reason:

> By ... admitting that there is a 'natural' division between women and men, we naturalise history, we assume that 'men' and 'women' *have always existed and will always exist*. Not only do we naturalise history, but also, consequently, we naturalise the social phenomena which express our oppression, making change impossible. (Ibid.)

Naturalist, biologist, essentialist narratives and explanations are, according to Wittig, part and parcel of a socially constructed ideological system that oppresses (and produces) 'women'. Whatever the intentions behind specific deployments of ideas of women's specialness or biology, the impact of their expression is to fix the possibilities of women's being to an *eternal* form, to freeze history by making a historical moment appear as eternal nature and to thus make feminist politics impossible. All such arguments need to be shattered so that the patriarchal social system can be 'unmasked' (*QF* collective, cited in: Jackson, 1995:14).

Wittig's critique of the biologism of 'early' feminism

Wittig explicitly defined her position in contrast to that of the 'early' feminists – by which she means European and North American feminists of the early twentieth century. In 'One Is Not Born a Woman' Wittig declared that when she and others named themselves 'feminists' (in 1971) they did so in order 'to identify a political link with an old movement'. It is as such 'this movement that we can put into question for the meaning it gave to feminism' (1993:105). And this Wittig does, positioning her materialist feminism as a progression from that other 'early' variety, as well as from those of her contemporaries who share in the 'early' feminists' failure to see the constructed, social, political and economic nature of the class woman – who fall back into the trap of the 'myth of woman' (Ibid.:106). Wittig identifies the 'early' feminists' commitment to the 'myth of women' – the idea that biology or nature is responsible for the distinction between men and women – as their big problem.

Not only does Wittig position the 'early' feminists as temporally prior in terms of their moment in a progressive history – they are also prior in their very *relation* to time. They are, she suggests, conservative, bound to eternal categories, blind to dynamism. And this relation to time – this stagnation – is (according to Wittig) the *effect* of their naturalising, biological conception of sex difference. They failed to see that the features of men and women were social rather than biological – even going 'so far as to adopt the Darwinist theory of evolution' (Wittig, 1993:106). 'Early' feminists could not see how to fight or what they should be fighting for, because they 'could never resolve the contradictions on the subject of nature/culture, woman/society' (Ibid.:105). They failed to see that 'woman' and 'man' are not eternal categories, they 'failed to regard history as a dynamic process' (Ibid.:106).

Wittig bases her claims upon an earlier article by Rosalind Rosenberg, 'In Search of Woman's Nature 1850–1920'. In this article Rosenberg tells something of an 'enlightenment story' about the development of an emancipatory conceptualisation of 'woman's nature'. Rosenberg claims that until the mid nineteenth century the 'popular conception of womanhood had always been fundamentally biological, even when given a spiritual elaboration' (1975:142). It is notable, therefore, that she sets out with an ahistorical approach to biological ideas and rationality. The article goes on to trace emancipatory progress in the conceptualisation of woman's nature, starting with 'early' feminists' insistence that women's nature is not in fact inferior to men's and then progressing to a growing awareness of the social, non-biological, and thus transformable character of sex difference. The end point in Rosenberg's account is in the 1920s when a handful of social psychologists 'finally' recognised that womanhood is actually a matter of custom. These psychologists, 'unlike Darwinism provided feminism with a strategy of reform' – a strategy that sadly was not taken up because most of the feminists 'still acted on faith that the expansion of the female sphere depended, in a fundamental way, on a belief in an innate female nature' (Ibid.:152).

'Early' feminists, according to Rosenberg, were limited in the scope of change that they sought because of their commitment to Darwinism and, with this, to the 'ancient belief in female uniqueness' (1975:142). Rosenberg characterises Darwinism as a conservative ideology in a fashion that appears to conflate modern evolutionary biology with the 'natural history' that preceded (and coexists with) it (as set out in Chapter 1; Foucault, 1970: esp. 263–79).

Rosenberg does acknowledge that Darwinism was helpful for the radical women of 'early' feminism. However, she effectively insulates the radicalism (and feminism) of the 'early' feminists from their Darwinism and biologism. Darwinism provided something like a necessary degree of conservatism and security for these women who were *otherwise* committed to radical transformation. She writes:

> Darwinism is most often described as an ideology in defence of the status quo...But Darwinism was just as important to the female advocates of social change at the end of the nineteenth century...For such women, Darwinism's biological determinism provided the reassurance they needed as an anchor of certainty in a time of social flux, an anchor that could protect them not only from critic's hostility, but just as importantly from their own uncertainty. (1975:142)

For Wittig and Rosenberg, biological thought is conflated with a belief in the necessity of the given social order and of women's nature in particular.

An immensely powerful conservative force is attributed to the belief in such ideas – that belief is posited as one of the major causes for the continuation of patriarchy. There is an apparent assumption that if only women knew that is was possible to change the nature of the sex distinction – if only the ideological plug were not in place – actual equality and liberation would automatically follow. This, in turn, implies something like a necessary force of liberation in the transformations of history – a force that is impeded by conservative counter-forces and ideological fallacies (a force somewhat ironically akin to that of 'evolution' as imagined in many philosophies of life). On a related theme, Butler notes a (paradoxical) commonality between Wittig's conception of emancipation and Marcuse's with his necessarily emancipatory Eros (Butler, 2004:32).

Rosenberg's and Wittig's account of 'early' feminism and Darwinism is a classic modernist tale in which the perspectives that the author agrees with are 'facts' that are 'realised', whilst alternative perspectives are 'faith' and 'belief' – those who believed in them being implicitly painted as immature versions of the authorial self. The positions are temporally hierarchised such that the eventual 'revelation' of the present position appears as the logical, if not *inevitable*, unfolding of the past. Wittig portrays the biological ideas and values of early feminists as the opposite of her own. She does not posit a dialectical opposition – in which the ideas she supports could be seen as an antagonistic development from the biologistic theses. It is more of a diametric style opposition – the different positions painted as stages on a linear development.

The problems with this as a general approach to history have been well rehearsed in the three decades since Rosenberg's and Wittig's papers were published. It would be churlish to hold Wittig and Rosenberg to account from the perspective of present historicity and scholarship. The problems with the materialist-feminist critique of 'early' feminists' biologism are of interest not because they 'show' that Wittig's and Rosenberg's readings of history were wrong but rather because that critique and its assumptions continue to shape the formation of feminist theory today. Normative choice in the feminist debates over ontology continues to refer to a choice between being and becoming. The distinction between being (born) and becoming (cultured) was compounded in critiques of biologism such as Wittig's and this distinction continues to inform feminist and sociological moral and political evaluations of ontology. By digging up alternative account of the history that lead us to that choice, by showing that Wittig's negating critique was not necessary and contrasting it with the findings of a positive critique, we might gain an insight into what is at *stake* in the formulation of our present values.

Dichotomies of being and becoming

A key operator in Wittig's critique of 'early' feminisms' biologism is the *temporalised logic of values* that she deploys. Wittig essentially posits that change and becoming are good and emancipatory whilst stasis is conservative, patriarchal and oppressive. She proceeds to characterise the biologism of the early feminists as conservative and contrasts this with a constructivist celebration of becoming. This leads to a problematic presentation of the relationship between biology and radicalism amongst the 'early' feminists, presenting the time of biology as that of eternity and preformation. This seems designed to draw a false opposition between constructivist and eugenicist ontologies in terms of a distinction between being and becoming, determination and contingency.

Wittig's opposition to biologism is a part of her political and ethical affirmation of the social and cultural in themselves, and of a perpetual work of transforming the self and others – becoming – in politics and art. This celebration of cultural change in itself is expressed in her 'anti-utopian' utopian novel *The Guérillères* (1972). *The Guérillères* is a fictional high point of the imagining of what Ruth Levitas calls 'process *as* utopia' (Levitas, 2003:146–7). Wittig does not simply celebrate process and transformation as a means to reaching a given telos – rather, perpetual change is itself the telos, the utopia, of her utopianism. There is a faint echo of Trotsky's call for permanent revolution (see Arendt 1968:384).

The text presents a fragmented narrative that throws up poetic images of moments in the lives of the Guérillères. Many of these moments are engaged in battle with some male adversaries. Many are fragments of stories or deliberations. Many are moments in which the Guérillères engage in battles with themselves, with their own language or ideology. Battles are not depicted from the perspective of winning or losing. Rather, the battles are transformative of all participants, including our protagonists. Although the narrative ends with an image of peace, this is not conclusive, for moments of peace have appeared before. It is a utopian work in which the transformation and battles by which the Guérillères become-other is never finished. The series of transformations – the becoming – that the Guérillères encounter in battle, in art, and in debate, is itself the utopia of Wittig's vision. Wittig's social constructivism is not, then, simply the denial of giveness and necessary permanence; it is also a political and ethical affirmation of the construction work *itself*, becoming, as its own end.

The distinction between being and becoming is a key organising principle in Wittig's values, one which she deploys throughout her critique of biologism; biology is associated with being, giveness, determinism and conservatism whilst culture and social structure are in the realm of becoming, transformation, creativity, radicalism and artistry. Becoming, transformation and creativity are treated as values and ends in and of themselves, without explicit differentiation being made between alternative types of becoming.

In order to maintain the logic of these oppositions, whilst situating her own position and biologism on opposing sides, Wittig seems to misread the character of biology and its values, masking the celebration of creativity, transformation, becoming and radicalism in biologistic thought itself. Such a strategy is apparent where, as we have seen, Wittig and Rosenberg carefully insulate the dynamism and progressivism of early feminists' vision and action from their commitment to Darwinism and biologistic beliefs – casting Darwinism as a conservative source of security for these *otherwise* radical women.

As we have seen above, this separation between biologism and 'early' feminists' commitment to transformation and progress, their radicalism, sits very uncomfortably with those feminists' own accounts of their motivations. Helene Stöcker, for example, stated of her own commitment to Darwinism that:

> If one believes in the eternal Becoming, in the flow of evolution, and holds struggle for the father of things, then one can only see

the moral task of humanity as seeking ever new, higher forms of morality. (cited in Weikart, 2004:132)

Indeed, it sits uncomfortably with the whole event of eugenics – which is to say, the politics of deliberate biological *transformation* – or with the project of modern biology, the science of *life*. As we have seen, leading feminists throughout Europe and America cited biological improvement as the very *raison d'être* of women's emancipation. Moreover, the idea of evolutionary life furnished feminists, with socialists and other reformers, with an immensely *dynamic* conception of history as well as a source of immanent authority for *radical* iconoclastic values.

Darwinism might not have furnished early feminists with strategies for change that Wittig and Rosenberg would, or any of us should, agree with. But biological reason and eugenicist policies certainly *did* provide strategies for change. When Stöcker affirms a belief in, and ethical commitment to, the 'eternal Becoming', she may be affirming a different radical ethics and politics of contingency to Wittig and Rosenberg; but she cannot seriously be held to be upholding conservatism or failing to see that history is a dynamic process born of conflict.

To be clear, I am not suggesting that eugenic feminism and materialist feminism did in fact support the *same* ontologies or values. Rather, I am suggesting that the way in which Wittig attempts to define the difference between them does not hold. Eugenic feminists and materialist feminists advocate different ethoses of becoming, different vitalisms, different radical feminisms. They do not represent alternative sides of a choice between stasis *or* transformation, being *or* becoming, conservatism *or* radicalism. The differences between Wittig's vitalist constructivism and Stöcker's vitalist biologism are great – but they are not, as Wittig suggests, the difference between tradition and politics. If the allure of biological discourse for 'early' feminists is to be described in terms of the allure of experience it is not, as Wittig implies, that of conservatism and tradition, but that of limit-experience, of becoming other, to which Wittig herself aspires.

There are, of course, very significant differences between Wittig's materialist-feminist ontologies and values and those of 'early' eugenicist feminists. These pertain not (as Wittig portrays it) to a division between being and becoming, but rather to alternative conceptions/compositions of the *field of contingency* in which becoming takes place – the choreography of dynamic relationality that comprises collective affective force (the consistency of the unitary living plurality, population life). Arguably the post-war turn to social constructivism and the wide-reaching critique of biologism marked less a rupture with biopolitics than a reconstruction of the biological field of contingency – of population-life and its racialised fragments. There is a case for interpreting second-wave feminist social constructivism as a renewal and reinvigoration of biopolitical feminism in the aftermath of fascist eugenics – renewing and reinvigorating the drive for empowerment, extending embodiment and developing governmentality, but also producing concepts of dynamically related cultural difference that could be utilised and co-opted for the justification of eliminative governmental strategies. We will now move briefly to a reading of the positivity of Wittig's own politics of construction and, in so doing, support the suggestion that there are strong parallels between the experiential economy of eugenicist feminism and Wittig's materialist feminism.

The positivity of Wittig's constructionism

In describing existent and past beliefs as ideologies to be transcended, Wittig, like eugenicist feminists before her, participates in a radical recomposition of forces: an expansion and transformation of the fields of virtual action. In particular education and ideas become radicalised as fields of intervention and embodiment. Education was also central to the practices of social hygiene and eugenics in which 'early' feminists had participated. The 'lady visitors' of early-twentieth-century London set out to educate working class women whilst European missionaries set off to 'civilise' the colonies (see Lewis, 2000:86). In the context of ideology-critique and social-constructivism, however, it is as though education has shifted from being *one tool* of governmental practice to become the *very life blood of the biopolitical population*. In the constructivist rubric education takes centre stage. It is education that carries the sins of one generation to the next. It is education that is the key site for radical intervention in the make-up and character of population life.

As Angelika Bammer argues, Wittig and her Guérillières redefined the meaning of 'action'; ' "action" can [for Wittig] mean any number of things: telling stories, building machines, playing games or fighting battles... political struggle can take the form of the invention of myths, resistance can take the form of sleeping' (Bammer, 1991:126). Wittig, writing with Sande Zeig in 1979, maintained that women had to learn to embrace violence (cited in Ibid.:127), in a minor repetition of the embrace of negative eugenics on the part of 'early' feminists. The violence that Wittig would have us embrace, however, is 'primarily symbolic', aimed at the ideological structures and myths that construct and legitimate the patriarchal order including, as the prime target, the 'myth of woman' (Ibid.).

Materialist feminism, like neo-liberalism, is addressed to a nature that has been artificially, socially, constructed through processes of education and socialisation. It is nonetheless *a nature* with its autogenetic processes, essences, dynamics, laws and forces. The constructed nature is that of gender, ethnicity and market conditions – nature in a world of 'Education, Education, Education!' Like neo-liberalism, materialist feminism is oriented towards an expansion, extension and deepening of political agency, now focused on the planes of ideational change, culture and education – re-educating, constructing, socialising, transcending illusion.

The possibilities for political and social action were intensified in the emergence of the constructivist rubric. In contrast to programmes effected through breeding, the effects of constructivist interventions could well ensue within the lifespan of individual generations (although any good sociologist would insist upon the intergenerational scale of most meaningful educational reform).

There is an intensification of contingency, certainly. But there is *not* an invention of contingency. Biologism – eugenics and all – was already a politics of time passing, a politics of contingency. Biologism is a *modern* political discourse, itself orientated towards a present passing away, celebrating the intensity of process, human powers of transformation and the experience of becoming or creativity. It is the investment of individual life and action within a 'quasi-transcendental', collective, dynamic life that made biologism such an appealing discourse – not, as might have been the case for eighteenth-century natural history, the affirmation and conservation of a preformed order within the world.

The missing matter of biopolitical racism

Another issue with Wittig's and Rosenberg's critiques of the biologism of 'early' feminism is that they make no mention of the race and class supremacism of the 'early' feminists, nor, indeed, of their investments in eugenicist ideas and projects.

For Rosenberg, the problem with Charlotte Perkins Gilman's commitment to biology is that it prevented her from comprehending the psychological transformations that would be necessary for the attainment of women's liberation – from understanding that differences between men and woman were not innate (Rosenberg, 1975:146). Following this, Wittig claims that the problem with 'early' feminists' biologism was that it restricted their capacity to conceive of history as dynamic processes, and of men and women as contingent classes (Wittig, 1993:106).

However attention to eugenic feminism and the racism, class-supremacism and support for negative eugenic practice with which the biologism of 'early' feminism was bound up, makes the limited and sympathetic nature of Rosenberg's and Wittig's critiques surprising. Gilman was heavily invested in supremacist ideas concerning racial evolution and abhorred mixed raced partnerships. Amongst the proposals of the early-twentieth-century eugenicist movement that she endorsed was compulsory sterilisation for 'the unfit' (Scharnhorst, 2000:68–9). Gilman's utopian imaginings included implicitly genocidal visions of racial purity, whilst her plans to liberate 'women' from the drudgery of domestic labour were dependent upon publicly enlisting the services of those who were supposedly 'better suited' to such work (Ibid.:70).

As such, and in contrast to Wittig and Rosenberg's suggestion that the problem with Gilman's biologism was that it prevented her from realising how to accomplish social change, we would be justified in concluding that the less Gilman understood about how to achieve her objectives *the better*. The failure to properly conceive of women and men as the same was the very *least* to be feared from her biological beliefs. If Wittig and Rosenberg were setting out to demonstrate the problematic nature of 'early' feminists' commitment to biologistic ideas it is strange that they made no mention of these issues.

The absence of a discussion of race and class supremacism cannot be put down to a lack of interest in these issues on the materialist feminists' part. As Bell argues, the issue of racism was central to second-wave feminist's concerns with biology. As we have seen above, this is manifest in Wittig's writing in the guise of references to African-American slavery. Wittig's interest in class oppression is in evidence throughout her discussions.

The absence might well be attributable to a straightforward ignorance. According to Anne Phillips, feminists in the 1970s 'knew little of what had gone before' (Phillips, n.d.). The historical record of eugenicist feminism has been more fully established in literature of the past two decades. It is certainly plausible, therefore, that Wittig and Rosenberg did not discuss the issues of 'early' feminists' supremacist beliefs and involvement with eugenics because they were simply unaware of them.

Be that as it may, the absence seems to reflect an unwillingness or inability to confront the biologism – the biologistic rationality and values – that was in fact adopted by many 'early' feminists such as Gilman, Sanger, Stöcker and Grand. To interrogate the supremacism of eugenicist feminist thought would have been to confront a discourse wherein racism and class supremacism were *central* to the conception of history as dynamic and contingent, as well as to strategies of progressive reform. It would, as such, have been to disrupt the dichotomous logic of values, and linear conception of emancipation in history, that Wittig's arguments develop.

Foucault, who developed a *positive critique* of the biologism and racism of the eugenicist era, highlighted the dynamic, inclusive and historic character of 'specifically modern' racism (Foucault, esp. 1978; 2003a; 2003b).[5] Likewise, Hannah Arendt characterised the racist ideology the informed late-nineteenth-century imperialism and twentieth-century fascism in the following terms:

> Ideologies are never interested in the miracle of being. They are his-
> torical, concerned with becoming and perishing, with the rise and
> fall of cultures, even if they try to explain history by some 'law of
> nature'. The word 'race' in racism does not signify any genuine curi-
> osity about the human races as a field for scientific exploration, but
> is the 'idea' by which the movement of history is explained as one
> consistent process ... Racism is the belief that there is a motion inher-
> ent in the very idea of race, just as deism is the belief that a motion is
> inherent in the very notion of God. (Arendt, 1968:469)

A processual racist logic is central to the eugenic strategies of social progress, as supported by eugenicist feminists. The fragmentation of the unitary living plurality – the population – into races and biologi- cally defined classes of people made it possible to conceive of a prac- tice wherein the collective life of society was promoted through the elimination, curtailment or improvement of given fragments (Foucault, 2003b:256–58).

It is, perhaps, no wonder that this modern racism, in which 'early' feminism was heavily involved, escaped the view of Wittig's and Rosenberg's critiques of biologism. That racism refuses to recede into a dusty past of patriarchal conservatism, from which feminists, with anti-racists, can be imagined as moving forward. It disrupts the bina- ries that Wittig's critique draws between the values of being and those of becoming. It demonstrates that radicalism and a conception of the social world as contingent are no guardians against the most problem- atic manifestations of biopolitics.

Problems with the negative critique of biopolitics

Wittig and Rosenberg engage in a negative critique of biologism. They seek to demonstrate that the ideas they are contesting are fallacies. They establish a dichotomous analysis in which one side of the 'debate' is established as the negation of the other. Whilst these arguments have considerable rhetorical appeal in their symmetry and simplicity – and certainly have had considerable strategic and ethical value – they resolutely do not succeed in grasping, let alone establishing a distance from, *biopolitical* rationality. The specificity of modern biological rationality is effaced in the dichotomous drawings; eugenic racism, the modern racism with which 'early' feminism was much involved, disappears from view. This is a specific instance of the general problems with negative critique as a strategy for addressing biopolitics.

The caricature widely endorsed by feminists, and sociologists more generally, of Darwinism as a conservative 'ideology of the status quo', could be seen as a part of the general modernist attempt to distance modernism from its inherent problems. In particular it could be seen as a part of the attempt to cast the Holocaust as outside of modernity – as some kind of anti-modern resurgence of barbarism (see Bauman, 1989).

In the instance of Wittig's and Rosenberg's accounts of 'early' feminism, the caricature of biologism as conservative serves to insulate the feminist and radical aspects of these women's commitments from their biologistic beliefs. Such an insulation might make it easy to affirm radicalism and feminism. It does also, however, efface important aspects of what feminism and radicalism have been and can be. Moreover, perhaps more importantly, it radically obscures the nature of supremacism and eugenic politics, painting the enemies of progressive feminism and anti-racism as dusty old conservatives, bent upon upholding a pre-given order of inequality in the world. The dynamism, progressivism, vitalism, modernism – and sometimes *feminism* – of eugenic supremacism simply do not come into the vista of this critique. This has the highly problematic implication of suggesting that all we have to do to ensure the non-complicity of our statements with regimes of racist fragmentation and biopolitical power is to avoid being dusty old conservatives. Unfortunately the politics of biopolitical racism is far more difficult and intractable than that.

Conclusion

The anti-determinist, anti-biologist, anti-essentialist, anti-racism of much feminist and sociological thought since the Second World War did not get a grip upon the progressive, processual variants of racism engendered in eugenicist and imperialist politics. The critique of Darwinism and biologism *as* preformism and fixism was inadequate to the task of exposing and opposing biological racism. It is not difficult to understand how such an inadequate critique could emerge from the strategy of opposition and negation embraced by Wittig and other exponents of 'ideology critique'. In Wittig's negating critique we can see how the effort to distinguish oneself from, and to condemn, a problematic political discourse can lead to the development of masks and disguises for that very discourse – making it in fact *easier* for the techniques and rationalities of that discourse to be repeated in newly configured fields. The condemnation of biology as a conservative 'ideology of the status quo' obscured the progressive, dynamic, processual character of biologism and of modern racism in the context of feminist critique. It might, as such, have had a part to play in facilitating repetitions of modern dynamic racism within the constructivist, culturalist rubrics that have followed. Foucault's positive, genealogical, insider critique of biopolitics takes us into some very uncomfortable places, attempting, in effect, to empathise with the positions that we most ardently oppose. But such discomfort is the price of a serious political economy of experiencing bodies that makes biopolitical racism properly *visible*, and that can, as such, foster demands and opportunities to generate and adopt alternatives.

The condemnation of biological politics has been effected in sociology and left-wing political discourse in large part through its presentation as *conservative,* upholding the hierarchies of the status quo. If its appeal for the subordinated has been explained at all in such critiques it has been in terms of the allure of conservative experience (a need for identity, epistemic security or tradition). This book, drawing out a particular reading of Foucault's positive critique of biopolitics and biomentality, has attempted to demonstrate that biologistic politics – the politics that *did* support horrifically pernicious supremacist practices of eugenics and imperialism at the start of the twentieth century and beyond – are 'progressivist' in the sense of being invested in the desire for transformation, historicism, contingency and radicalism. This chapter has highlighted the close association between biologism and feminism in the eugenicist era, giving empirical weight to that analysis.

The extension of biopolitical governmentality to some extent required and enabled an empowerment of women and a making public of the private sphere. A modernist politics of transformation, contingency and empowerment was immanent to biopolitical governmentality. We *can* understand the appeal of biopolitics for 'early' feminist women in terms of the allure of experience. The desired and authoritative experience, however, is *not* about an embedding in a fixed order, stability and security (as some, such as Wittig and Rosenberg have supposed). The experience that grants biopolitical discourse, or biopolitical power, its 'hold' upon subjects is the experience of breaking limits, of transformation, power and self-transcendence, experience as life, *modern* experience, which is a similar kind of experience to that promised by anti-biologistic constructivism and processual utopianism.

This does not imply that socially-constructivist and contemporary processualist discourses are necessarily or probably racist. But it *does* imply that when confronting the problem of racism today (a racism that might be primarily articulated in terms of cultural differentiations rather than the biological) we should not fall back on an assumption that so long as our discourse in neither reifying nor exclusionary (or 'Othering') we are safe from participation in the experiential economies of racism. In order to avoid essentialism and the destructive governmental practices that it makes possible, it is not enough to avoid reification or determinism, nor is it enough to insist (again) upon the reality, desirability and ethos of becoming – be that the becoming of a performative discourse, the becoming of bodies, or the becoming of 'life itself'. If eugenicist feminism was invested in philosophies of becoming-other and if biopolitical racism was a part of the technology of such transformation then Wittig's strategy of opposing determinism, preformism and conservatism was wholly inadequate – at least as a strategy to ward off biopolitical racism and supremacist-specificiation. Racism and supremacism can not be easily situated on one side of the lines of opposition between constructivism and preformism, progressivism and conservativism, contingency and determination, or becoming and being. To get to grips with the peculiarly biopolitical, modern formation of racism – a form that has been immanent to feminism and socialism in the past (at the least) – we have to understand the dangers of that are *immanent* to politics of process, transformation and life. We will fall short if our primary critical strategy is to demonstrate the complicity of various 'ideologies' with determinism, preformism or stasis.

Conclusion

Foucault's account of biopolitics, as interpreted in these pages, is focused upon *experience* – a cartography of biopolitical, or bio-mental, 'foyers' (hearths) of experience and of the plays of perception, affect, desire, and sense of reality that they make at home. This account does not reveal the 'truth of an age' or 'a civilisation', nor is it a total theory of modern politics or power. It is an attempt to capture the difference that is made to some political games of authority, persuasion and domination by a set of innovations in knowledge. It is not simply a history of medicine, bio-industry or science – knowledge is the general organisations of what is seeable, sayable, affectable. Foucault's positive-critique of biopolitics is a work to make more visible, and more contestable, the *conditions* of certain ways of being political. 'Biological knowledge' or 'bio-mentality' names an aesthetics, a structuring of experience, a (sufficient) condition of subjectivity in all its autonomy, vitality and purpose, its meaningfulness, affective capacity and depth. Bio-mentality is a production of what counts and how it is counted, a mattering (making what matter). Biopolitics describes important innovations in the structuring of experience, but it is not *the story* of modern government. History is manifold plurality. Arts of government, power and legitimisation plug in to a plethora of experiential truth- and other-games. Foucault's analysis of biopolitics provides some tools for understanding (and combating) what is going on when politicians, activists and other artists play upon the heart strings of security, endless growth, infection, creativity, vitality and freedom; population, class or nation (etcetera) – pointing out some of the conditions upon which such games depend and giving us a handle to grasp some of the rules and technologies.

For Foucault 'experience' is limit-experience; an encounter with the limits of the possible. It is a process of desubjectivation – a getting away from the self – most readily associated with transgression, madness and death. Indeed, Foucault argues that the experience of biological life first took place whilst cutting up corpses; in Bichat's big eye that had stared at death. To experience is to experiment; to interrogate and being interrogated by the world. Experience means being a part of the universe, of its teeming transforming capacities – going beyond self, moving outside of singular subjectivity. Self-transcendence. Experience is a mode of contact with forces beyond the self (not only ones own self, but also selfhood, subjectivity, in general); a becoming objective. Experience generates a sense of reality, authenticity and epistemic security.

Bio-mentality (biological knowledge) constituted a radical innovation in the conditions of experience. Biological knowledges, modern sexuality and public heath discourse enfold individual bodies into the collective life of populations. Biological life creates a new dimensionality to present existence and it folds present mundane bodies and their health into this self-transcending, intensive domain. Bodies spill over into each other in biological perception, in a way that is simply not true of bodies as apprehended in classical knowledge. Biopolitical embodiment enfolds experience into the everyday, the everywhere and everyone. Experience, once the preserve of the momentous moment, the aristocrat and the mystic, becomes entwined with the mundane mortal processes of family, sexuality and health. When life becomes the limit (the outside and the objective) authority, (subjective power) proliferates and is pluralised.

The concept of population is, then, crucial to the analysis of biopolitical experience. Population – or an immanent plane of trans-organic, trans-personal connectivity, intensity and growth – is the *necessary* condition of life as biopolitics knows it however 'alternative' the biopolitics in question might be. Population is a condition of life as self-regulating, creative, evolutionary, radical force. It is not *living* but population-life that is attributable with those spectacular powers that engorge the time of the now and make individuals feel themselves capable of ever inventing the universe anew. It is in the milieu of population that we get beyond ourselves, and this self-transcending movement that makes bio-mentality such a subjectively successful, enriching, discourse – folding transcendence, excess, externality, *being collective*, within the impulses and affects (influences) of organic bodies and mortal lives.

Biopolitical authority is authority that obtains from having experienced (population) life; from having battled with the limits of life. Biopolitical authority is not the rule of scientific truth – nor of a despotic, totalising life force, collectivisation or sovereign. It is, rather, the power and attraction that congeals around a diversity of performances and manifestations of experiencing life. To be biopolitically authoritative is to mediate experience of life, to be a conduit to the force by which life (objectivity) pushes back. To know life, to make life manifest, to make a promise that life is real ... to provide a link to life is to generate biopolitical authority. Of course biologists and doctors with their tests and interrogations are authoritative in the biopolitical worlds. But so too are those that have encountered the edges of life – moved close to death, created new lives. Crucially, markets, with their capacity to test imaginings against the free flux of life in its 'natural laws', bear immense authority. Markets are mechanisms by which liberal and neoliberal governments test their governmental theories against the unfettered forces of life.

Foucault's writings on biopolitics have fed into critical work upon the ways that medicine, security and capital makes people's life *object* of power – helping to identify processes of exploitation, normalisation and elimination of life. In an age of emergency, rampant militarisation, intensified capitalisation, consolidation of elite privilege, and reinvigorated racism, such processes have rightly been at the forefront of a great many critical minds. But Foucault's own concern with the power-knowledge of life was bound up with a project to comprehend the production, not of powers' objects, but of *subjects*; capacities for experience, evaluation and action. The archaeologist digs up the positive-unconscious that constitutes ways of knowing, of looking, of experiencing. The genealogist interprets such structures as moments in the multiple history of bodies – where bodies are capacities to affect (as in 'to influence') and to be affected; circuits, pathways and intensities of force. These practices of positive-critique engender a kind of ethical anthropology of present power – an empathetic, pragmatic, interpretation of the forms of embodiment, culture and reason that sustain certain forms of governance.

The positive-critique of biopolitical experience is not about identifying and denouncing objectifications or destructions of life. Nonetheless it certainly is a *critical*, ethical pursuit. To write the positive critique of biopolitics is to cultivate an aesthetics of population – a conscious creative work upon the conditions (the limits) of life. It is a critical, ethical, practice because it develops awareness of where we come from, intensifying reflexivity, aestheticising (making subject to conscious constructive practice) our strivings for truth, progress, significance and connection. It is an attempt to become more responsible for our self-production, more conscious of the causes of our desires, more adept at human engineering. The positive-critique of biopolitical experience is not a moral condemnation of the world, power or politics; but it is a practical endeavour to become more active, affirmative and constructive – more conscious of and responsible for the ways that we are empowered.

Past biological politics continues to shape present politics, social ontology and experience, not only through its persistence and repetition in new (and old) forms of population politics, but also as a spectre of the past that haunts present social and political theory, defining what we no longer are or what our enemy is. When biopolitics appears in a dramatised caricature (as thanato-political or conservative or homogenising) it is too easy to imagine ourselves as radically different, too tempting to assume that we can easily extract ourselves from biopolitical history and from the productions and power in which our agency has been constructed.

Foucault's thinking, as interpreted here, illuminates a mundane, positive, history of biopolitics: a biopolitics that extends far beyond concentration camps and forced sterilisation programmes; a biopolitics that is necessarily immensely plural, creating limit-experience, intensity and contingency, as a crucial aspect of its own grip upon power (inventing, inventing and reinventing as a matter of necessary course); and a biopolitics that enables the everyday world-building connections (of families, communities, economies and endeavour) that produce significance and intelligibility in a radically finite, transforming, contingent, fragile present. This account shifts attention away from thanatopolitics and concentration camps towards the mundane and even 'beautiful' investments of bodies, capacities and life. Foucault's account of biopolitics focuses less on processes of objectification and dehumanisation than it does on the trans-organic connections and limit-experiences through which modern *subjectivity* is produced. It is, however, no less critical. The positive critique of biopolitics does not wipe concentration camps, internment camps, sterilisation programmes, or ethnic genocide from the annuals of biopolitical history. Rather it pays attention to the mundane, multiple, positive processes of production, embodiment and experience that constitute the rationality, viscosity and affective force of biopolitical power, which is to say the conditions of biopolitics (in its deathly lurches and otherwise).

The positive critique of biopolitics is a tool for working upon the conditions of present economy. Understanding the allure and mundane production of biopolitical power, embodiment and discourse helps us to pluralise the present, to loosen the grip of political discourses and to make our own habituated investments strange. The more we describe the positive production of the power that we inhabit, that produces our embodiment, investments and experience, the more capacity there is to take hold of and re-appropriate that power, to reinvent it, redirect it; a practical attitude that seeks an actively critical approach to power in the present – not moral condemnation, dramatisation and regret ... an aesthetics of existence.

This ethical work upon the limits of our existence (critically comprehending our subjectification) is not about navel gazing, soul searching, individualised practice. We become subject through practice in common technological contexts. Biological type knowledge, or 'bio-mentality', is a crucial character in the conditioning of our subjectification; we are made subjective (sometimes) through our technological composition as biological life. Making this technology more visible gives us a more critical relationship to ourselves as the agents and audiences of biopolitical discourse. Positive-critique of biopolitics is, then, a practice of ethics – an aesthetics of existence that is not about the individualised pursuit of 'a beautiful life', but a collaborative work to make visibly created, creating and contestable (subject to work) the conditions that collectively constitute us *as life* – a creative practice of becoming together.

Foucault's methods, genealogy and archaeology, are often associated with anti-essentialist affirmation of contingency – denaturalising discourse, demonstrating that things could be otherwise. Certainly Foucault's work is engaged in a loosening of present claims. The freedom that positive-critique aims to foster, however, is not simply a (negatively defined) 'freedom from determination'; the aim is not to escape power, nor to celebrate contingency in and of itself. The affirmation of contingency qua contingency is a figuration of the faith in linear progress and the self-generated subject against which Foucault rightly railed (and which continues to sustain some of the most problematic and racist modes of biopolitical government). The point of positive-critique is not simply to know and assert that thing could, in theory, be otherwise – it is to know more about how things have become what they *in fact* are, to develop comprehension of our determination, to get to grips with the technologies of our production. It is understanding determination (pragmatic analysis) of the processes and limitations of our present construction that enhances capacities to engender real transformation and political action.

Describing biopolitical technologies in their positivity can make their subjective grip, and its workings, more visible and open up new flanks in the battles against intolerable governance. For example, Foucault conceptualised 'biopolitical racism' and described its capacity to intensify the transformability of population life and to make (directly and indirectly) murderous practices experienceable as maximisation, purification and growth of life. This suggests (as have many other thinkers) that confronting racism is of utmost priority in *all* efforts to develop good governance in our biopolitical world – anti-racism is no 'single issue' pertaining to particular interest groups, it is fundamental to all struggle against domination, militarisation and impoverishment. Foucault highlights figurations of racism that are invisible or obscured in the narratives of negative-critique; flagging up forms of racism that are 'progressive', inclusive and invested in contingency. This racism, which Foucault argues is inscribed at the heart of technologies of modern power, is not the 'ethnic' racism (of 'them and us') that, he suggests, has probably always existed in some shape or form. It is a specifically modern racism, operationalising specifically dynamic biological difference, and it can be activated with respect to all sorts of social problematics and fragmentations, including class, health, crime and religion in addition to phenotype or ethnicity.

Throughout this book I have stressed the distance between biopolitical racism and 'the ideology of the status quo' and argued that if the latter does characterise an ontology of nature it is that of classical physiology or 'natural history', not biology. I have attempted to demonstrate something of the allure and appeal of biopolitical racism, noting the use of racism in technologies of power/knowledge to politicise and processualise the present, making the present population more subjectable to the agency of apparently 'progressive', 'evolutionary' or 'vitalising' political action. There is a dimensionality to biopolitical racism. Not only is it many faceted, multidimensional, plugging into all manner of political discourses, movements and technologies (including of progressive politics). It is also *productive of dimensionality*, producing historicity, present depth and limit-experience. Obviously the point of this depiction of the positivity of racism is not to celebrate, condone or excuse such racism. Foucault's positive-critique in fact suggests that biopolitical racism is more problematic and more *dangerous* than the negative accounts of biopolitical racism (as productions of enmity, Othering, or as ideology of the status quo) imply. Foucault's analysis suggests that biopolitical racism enables a positive vitalising value and drive to be attributed to the politically and physically murderous functions of the state and of political movements. The positive, 'progressivist', 'appealing' character of biopolitical racism means that modern racism is more intractable, harder to resist and more viscous than anti-racist sociology has often assumed. It is easy to assert that we should avoid producing relations of enmity. It is more difficult to contend that we should avoid a politics of 'progress' or a work towards health and vitality.

Foucault does not engage in the kind of critical practice that denounces racism or massacres as morally wrong – the wrongness of such things goes without philosophers saying! Nonetheless describing such racism in terms of its positivity, attempting to comprehend and make visible the ways that it works subjectively (without recourse to the supposed depravity, idiocy or passivity of its audience) is a vital ethical practice. Sounding notes of resonance between the politics we most wish to do away with and our *own* ambitions fosters a critical awareness and power with respect to the causes of our desires, illuminating our investments in biopolitical power. Comprehending better (some of) the subjective technologies that facilitate exploitation and securitisation can help us to forge the tools of alternatives. Moreover, recasting discourses of enmity, 'good and evil' or 'life and death' in more mundane technological terms can deflate the soaring rhetoric in which racism thrives – slow the tempo, cultivate vocabularies and time for cooperative practice in which energies can be refigured and reinvested.

Another issue in this book has been the actually and potentially *cultural* character of biopolitics and biopolitical racism. Trans-organic embodiment, population life and population fragmentation are not the sole preserve of organic connections and flows: biopolitics can be 'culturalist' rather than 'biologistic'. What is most essential to the political economy of experience (typified in the politics of transorganic life) is not the organic but the *trans*. Technologies concerning *cultural* connectivity, capacity and value can have the characteristics of biopolitical discourse. Culturalist biopolitics might pertain to technologies and discourses of creativity, cultural difference or development, education and 'ways of life'. Nineteenth century conceptions of life, evolution and biological differentiation were intertwined with notions concerning culture and 'civilisation'. Contemporary politics of cultural value, vitality and difference might operationalise the technologies of the biopolitical economy of experience. The dangerous play of biological type relations might be activated along lines of cultural fragmentation – operationalising cultural differences in divisions of class, health, ethnicity, religion and sexuality. Recent decades have seen a shift in salience from biological to cultural racisms in many important arenas of governance. Fears around civilisation, infectious fundamentalism and cultural threats animate international security discourse whilst an international hierarchy of authority-to-act draws fragmentations in terms of cultural capacity, education and 'human development'. Matters of cultural difference (between ethnic groups, religions and classes) have gained a phenomenal saliency and affective force in national and European policy and public debates.

Normative descriptions of cultural difference do not simply give voice to prejudices or enable people to construct 'identities', they constitute specific organisations of visibilities, embodiment and powers; objects and capacities of governance; paths of influence and action; ways of caring for life. A meaningful critique of new cultural formations of biopolitical racism needs to take seriously the positive, productive and transformative force of such discourse. We should not assume that racist, supremacist and specifying discourse is necessarily reifying, deterministic or exclusionary. Discourses that recognise and affirm the dynamic, contingent and porous nature of culture and ethnicity might (intentionally or not) participate in technologies of biopolitical racism. We need to be alert to the dangers of dynamic, inclusive discourses of fragmentation – discourses such as that of 'development', 'enlightenment', and indeed 'emancipation'. Whilst we might be rightly bound to processual ontologies and a comprehension of bodies in terms of their becoming, the valorisation of contingency, vitality or becoming *in and of itself* is highly problematic.

Finally, the positive-critique of biopolitical experience problematises the contemporary ideas of a 'politics of life itself' beyond population, of 'biopolitics without biopower', and of politics as 'withdrawal from life'. Noticing the centrality of contingency, creativity, care and culture to biopolitical productions in the eugenicist past should make us more wary of vitalist ethics and aesthetics – at the least we can conclude that it is important to choose our vitalism carefully. At the same time, following genealogies of production of life as experience, tracing contours of technologies and knowledges that *make* biological embodiment, should leave us alert to the immanence of life to power, knowledge and technology. There can be no politics of creative life that is not also a politics of producing population, collective embodiment, that constitutes the limits through which self-transcendence takes place.

If life is produced (as a self-generating, self-ordering force of creativity) in history and in power, precisely in the production of trans-organic embodiment or 'population life', then there is no life outside of those forces of production, no life without the production of trans-organic population and investments of force. We might reinvent and redirect power and life but we cannot separate life from power. The positive critique of biopolitics, the account of life as experience produced through historical regimes of knowledge and technology, does not lead to a politics or ethics of 'free', 'real' or 'authentic' life. If it suggests a vitalist or new biopolitical endeavour it is that of opening up the present and pluralizing formations of life as experience. This is not about liberating, enhancing, harnessing or deferring judgement to a particular type of life, and it is not a matter of getting our life philosophy right. Rather it is a work upon the *context* and technologies of *production*: a political economy of experience; a politics of *conditions*.

Instead of a politics of 'life itself' or 'corporeality' the positive critique of biopolitical experience directs us to work upon the technological and political production of trans, of planes of limit experience. The multiplication and production of creative politics does not mean making norms of life. It means complicated, pragmatic, specific work of producing and augmenting relations: nodes of power-transfer through which affect and influence can flow and transform; generations of embodiment. This would not be a work upon power in the name of a life (that might escape it) but a work upon the limits of life – the fields of visibility, the imaginations, the embodiments, the technologies of connection and relation in which limit-experience is produced as virtual – in the name of specific augmentations and redistributions of power. We will foster our own biopolitical agenda and authority by developing *conditions* of life – culture, economy, biology; exchange, infection, mutual-influence. Our aim should be to cultivate technologies for the intensification of life that cast biopolitical racism and securitisation into shadow.

Withdrawing from life momentarily – putting ones life on the line, demonstrating that the will to life does indeed have conditions... such courageous acts have enormous politicising potential in biopolitical worlds, bringing the conditions of life into question, breaking the spell of a set securitising specification. Such courage composes a radically creative space in which new worlds become possible, the moment that Foucault calls 'political spirituality', an incantation of what Arendt calls 'the public'. But such withdrawal of life is a momentary aestheticising action – an event of de-territorialisation that opens contingency to the inscription of the next 'state machine' as readily as the next revolution.[1] Practicing the aesthetics (rather than simply the aestheticisation) of biopolitical life means sustained, mundane, practical work of governing, building, playing and performing our interconnection; embracing influence, energy and responsibility; working out, as best we can, how to do power well. Combating racist biopolitics means embracing, not running away from, biological life and its power: working deliberately upon the generation of lifes' conditions; finding or augmenting non-racist mechanisms of population-intensification.

If we are striving for creative politics, for a proliferation of life (of the capacity to err) then our sights should be trained on the multiple character of population, of connected difference, or 'trans', from which vitalities emerge as experience in the production and passing of limits. Such a politics is not an affirmation of a singular life that has emerged in a particular scientific or philosophical configuration of connection. Life itself is plural (lifes not lives), they are historical and fully immanent to power. There is no politics of life itself or biopolitics that is not also a politics of (bio) power: a political work upon the creation, negotiation and specification of our connectivity, our mutual influence, our biopolitical trans.

Notes

Introduction

Significant parts of the sub-section 'Benjamin, vitalism and the transformation of modern experience' appeared in my earlier article 'Destroying Duration: The Critical Situation of Bergsonism in Benjamin's Analysis of Modern Experience' *Theory, Culture & Society* vol. 25 no .3 (2008) and are reproduced here with the permission of SAGE.

1. 'Vitalist' – which is to say rationalities, motivations and ethics that conceive of the material and/or human world through a (more or less) 'anti-reductive' perspective, as comprised of self-generating and self-regulating, self-assessing/justifying, auto-normative, autonomous forces, and which celebrate or defer to life, vitality or creativity as the inherent value and source of verification (of truth and falsity production) in the world.
2. See Deleuze (1991). For critical accounts of the metaphysical, ahistorical character of Bergson's and Deleuze's philosophy see Osborne (2003a) and Hallward (2006).
3. *Erfahrung* is translated into English as 'experience'. In its German usage, however, it has the more specific sense of experience that is accumulated over time. For Benjamin *Erfahrung* is defined in relation to *Erlebnis*, the moment of lived experience.
4. On the conception of politics as speech and action in the public, see Arendt (1998: chapters 2 and 4; *esp.* 22–37 and 199–212).
5. It is this ethos – I would suggest – that leaves readers of Foucault and Deleuze frequently frustrated by the lack of references, and 'failures' to explain who or what is under critique and contestation.

1 Escaping the Laws of Being: The Character of the 'Bio' in Foucault's Genealogies of Biology and Biopolitics

1. Exceptions to that tendency include Melinda Cooper's Life As Surplus (2008), as well as works that, following the lead of Gilles Deleuze, draw together the issues of liberal subjectification and the emergence of biological life (with labour and language) (Deleuze, 1988; Dean, 1996).
2. On some of the biopolitical consequences of the molecularisation of biology see Rose (2007) and Rabinow (1999). For a schematic overview of the history of molecular biology see Rheinberger (2008).

2 Incorporation: Foucault on the Co-Constitution of Modern Embodiment, Experience and Politics

1. 'Sexuality' in Foucault refers to discourses and techniques surrounding sexual practice and desires, authority and expertise about sexuality, as well as the intensities that are produced in bodies through them.

2. Foucault's argument about the role of sex in masking the role of power resonates with Bourdieu's concept of 'symbolic violence', where a hierarchical
process of social reproduction produces effects that appear to be natural and
thus legitimise the constructed order, masking the processes of social reproduction (Bourdieu & Passeron, 1977).

3. This reflects the historical situation of the emergence of discourses of sexuality at a moment when Lamarckian theories of evolution exercised a major
influence (see Jacob, 1973: 217).

3   Christianity, Process and Positive Critique: Rethinking the Resonance between Foucault and Arendt, against Agamben

This chapter was first published as 'Foucault's and Arendt's "insider view"
of biopolitics: a critique of Agamben' in *History of the Human Sciences* vol. 23
(Blencowe, 2010).

1. For an extended critique of the overly abstracted nature of Agamben's concepts of biopower and sovereignty in contrast to Foucault's, see Rose and
Rabinow's 'Thoughts on the Concept of Biopower Today' (2003).

2. On the 'emergence' of population and with it the human sciences that analyse man as living being and species (as speaking subject and working individual) Foucault said: 'A constant interplay between techniques of power
and their object gradually carves out in reality, as a field of reality, population and its specific phenomena. A whole series of objects were made visible
for possible forms of knowledge on the constitution of the population as the
correlate of techniques of power. In turn, because these forms of knowledge
constantly carve out new objects, the population could be formed, continue
and remain as the privileged correlate of modern mechanisms of power'
(2007: 79).

3. With no argumentation or evidence Agamben claims that the subjectifications that Foucault identified in antiquity become objectifications in modernity (Agamben, 1998: 119).

4. For a considerably more extended discussion of the relationship between
Foucault and Agamben, which engages more extensively with Agamben's
literature and draws out similar themes to those raised here, see Ojakangas
(2005). Ojakangas argues that Agamben misrepresents Foucault's position on biopolitics because he obscures the historical specificity, as well
as the positivity, of biopolitics. For Ojakangas, however, that positivity is
tied exclusively to the agency of care, such that biopolitical rationality can
be equated with the *cura mata*. The problems with such a politics are
akin to those with an overbearing mother; biopolitics might be suffocat-
ing. This contrasts with the position that I am developing in this paper
insofar as I am arguing that the positivity of biopolitics also includes a
perpetual compulsion towards, and allure of, expanding forces and processuality. The point upon which this would bring Ojakangas and myself into
tension is with respect to the relationship between biopolitical values and
modern, biopolitical, racism and its relationship to modern genocide. For
Ojakangas, biopolitics, given biopolitical racism, can *justify* the exercise of

the sovereign power to kill, but that is all. It could not, as it were, *encourage* governance through elimination. In contrast I would argue that given modern racism the biopolitical injunction to maximise life and the perpetual biopolitical need for transformative can indeed be exercised through processes of political and physical elimination in which the quality of collective life is 'improved'. Her insight into such dynamics, especially in the context of totalitarianism, is one of Arendt's most significant contributions to the present interpretation of theories of biopolitics. It is an insight that I am convinced Arendt shares with Foucault, even if he does not express it so well.

5. Braun does also point to the zoëification of life as a major aspect of biopolitics, but that is not a major aspect of her article and is less relevant to our present considerations.

6. Braun references a conference paper presented in 2005 by N. Gerodetti, 'Biopolitics, Eugenics and the Use of History'.

7. On ethical thinking see Bernauer (1992: 268–72).

4 'Post-Population' or 'Cultural' Biopolitics? Rethinking Foucault's Concepts Today, against Nikolas Rose

1. Rabinow also cites Gilles Deleuze's 'Appendix on the Death of Man and Superman' in *Foucault* as an inspiration, where Deleuze suggests that since the nineteenth century we have left the 'Man form' associated with discipline and biopolitics behind, and moved towards a new field of the 'afterman' in which finitude and empiricity give way to a play of forces and forms; the superfold (Rabinow 1999: 407; Delezue, 1988: 102–10).

2. The ostensive rejection of eugenicist ideas did not prevent the governments of the United States, many European countries (especially the Scandinavian countries, the UK and Germany) and, in recent decades, Japan, from quietly supporting the continuation of eugenic policies beyond their borders in many 'developing' countries where western funded 'aid' programmes have frequently been responsible for encouraging sterilisation and fertility control 'population' programmes (Grimes, 1998: 383). The US Agency for International Development (US AID) was, for example, instrumental in stimulating fertility control activities in the Philippines in the 1960s and in mobilising the Philippine government to adopt an explicit population policy in 1970 (Warwick, 1982: 16). A particularly controversial episode in the 1980s and 1990s was that of trials of *Norplant*, a contraceptive implant, in various non-western societies including Brazil, Bangladesh and Haiti. These trials, which were funded by US AID, involved extremely dubious practices, with women being misinformed as to the experimental status of the drug and with some even being refused removal when the side effects became unbearable (Barroso & Corrêa, 1995; *Horizon*, 1995).

3. I am grateful to Victoria Margree for suggesting the national curriculum as an example of culturalist biopolitics in the questions following my conference paper 'Meaningfulness and Foundationless Power: Population, Process and Life' *Power: Dynamics Forms and Consequences* University of Tampere, 8 September 2008.

5   Eternally Becoming: Feminism, Race, Contingency and the Critique of Biopolitics

A truncated version of this chapter was first published as 'Biology, Contingency and the Problema of Racism in Feminist Discourse' in *Theory, Culture & Society* 28(3), 2011 and is reproduced here with the permission of SAGE.

1. By 'supremacist-specification' I mean hierarchical classificatory schemas that define, judge and act upon people in terms of 'their' classifications and which differentially value the life of different fragments or classes. As well as Foucault's concept of 'biopolitical racism' (which is somewhat confusing in Foucault, in that it refers to the politics of fragmentations and classifications of sexuality, class and health as well as ethnicity), I am drawing on Peter Hallward's use of the term 'the specified' (Hallward, 2001: 40; Foucault, 2003: 254). Hallward suggests 'specification' as a kind of general term to capture the dominating and 'emancipatory' processes of essentialism, rendering passive, rendering group, rendering object (which are resisted through movements from the specified to the Specific). In the specification of the specified 'what counts is the compliance of actors with a presumed nature, and the consequent supervision of the relative authenticity of that compliance' (2001: 40).
2. The term 'early' feminism is extremely problematic, given that there has never been a time in which there was not feminism of some description and that the term implies a misleading singularity across the plethora of different feminisms (for which, in fact, no one thing could be called the early variety). However all terms with which one can refer to periods of feminism are problematic and, contested. I have chosen this term because it is the one deployed by the materialist feminists that I will discuss in the larger part of this chapter and because, whilst inaccurate with respect to the history of the world, there is a certain accuracy in referring to the period of the suffragette movement as early biopolitical feminism. It would be more accurate, but more longwinded, to refer to 'turn of the twentieth century western feminism'.
3. Her *The Heavenly Twins* (1893) sold 20,000 copies in Britain within a few weeks and more than five times as many in the United States (Richardson, 2000:45).
4. It should be noted that Walkowitz developed a more complex, pluralist, picture of the modernity of early twentieth century British feminists, informed by post-structuralist debates, in her 1992 *City of Dreadful Delight*.
5. Both Arendt and Foucault make the claim that racism did not emerge until the nineteenth century but Foucault goes on to differentiate between much older traditions of 'ordinary racism' and specifically 'modern racism' (Arendt, 1968: 158–9; Foucault, 2003b: 258).

Conclusion

1. On de-territorialisation and state machines see Deleuze & Gurattri (1988).

Bibliography

1911census.co.uk (n.d) 'About the 1911 Census', *Find-my-past.co.uk; The National Archives* www.1911census.co.uk/Content/default.aspx?r=24, last visited December 2009.

Abu-Lughod, L. (2002) 'Do Muslim women really need saving? Anthropological reflections on cultural relativism and its others', *American Anthropologist* 104(3):783–90.

Agamben, G. (1998) *Homo Sacer: Sovereign Power and Bare Life*, Stanford: Stanford University Press.

Agamben, G. (2004) *The Open: Man and Animal*, Stanford: Stanford University Press.

Agamben, G. (2005) *State of Exception*, Chicago: University of California Press.

Arendt, H. (1968) [1948] *The Origins of Totalitarianism*, San Diego: Harcourt.

Arendt, H. (1993) *Between Past and Future*, New York: Penguin.

Arendt, H. (1998) [1958] *The Human Condition*, Chicago, The University of Chicago Press.

Arnold, M. (1869) *Culture and Anarchy: An Essay in Political and Social Criticism*, London: Thomas Nelson & Sons.

Bammer, A. (1991) *Partial Visions: Feminism and Utopianism in the 1970s*, London: Routledge.

Barroso, C. & Correa, S. (1995) 'Public Servants, Professionals, and Feminists: The Politics of Contraceptive Research in Brazil' in F. D. Ginsburg & R. Rapp (eds.), *Conceiving the New World Order: The Global Politics of Reproduction*, New York; London: Harvester Wheatsheaf.

Bauman, Z. (1989) *Modernity and the Holocaust*, Cambridge: Polity.

Bauman, Z. (1995) *Life in Fragments: Essays in Postmodern Morality*. Oxford: Blackwell.

Beard, J. (2006) *The Political Economy of Desire: International Law, Development and the Nation State*, London: Routledge, Cavendish.

Beauvoir, S. (1960) *The Second Sex*, H. M. Parshley (trans.) London: Landsborough Publications.

Bell, V. (1996) 'The Promise of Liberalism and the Performance of Freedom' in Barry, Osborne & Rose (eds.) *Foucault and Political Reason: Liberalism, Neo-Liberalism and Rationalities of Government*, London: UCL Press.

Bell, V. (1999) *Feminist Imagination: Genealogies in Feminist Theory*. London: Sage.

Benjamin, W. (1999a) [1936] 'The Work of Art in the Age of Mechanical Reproduction', *Illuminations*. London: Pimlico.

Benjamin, W. (1999b) [1939] 'On Some Motifs in Baudelaire', *Illuminations*. London: Pimlico.

Benjamin, W. (2002) [1936] 'The Work of Art in the Age of Its Technological Reproducibility', *Walter Benjamin Selected Works Volume 3*. London: Harvard University Press.

Bergson, H. (1910) *Time and Free Will*. London: Sonnenschein.

Bergson, H. (1935) [1932] *Two Sources of Morality and Religion*. London: Macmillan.

Bergson, H. (1998) [1911] *Creative Evolution*. New York: Dover Publications.

Bernauer, J. (1992) 'Beyond Life and Death: On Foucault's Post-Auschwitz Ethic' in Armstrong (trans.; ed.) *Michel Foucault Philosopher*, New York: Routledge.

Beveridge, W. H. B. B. (1942) *Social Insurance and Allied Services. The Beveridge Report in Brief*. [S.l.]: H.M.S.O.

Birke, L. I. A. (2000) *Feminism and the Biological Body*. New Brunswick, NJ: Rutgers University Press.

Blencowe, C. (2008) 'Destroying Duration – The Critical Situation of Bergsonism in Benjamin's Analysis of Modern Experience', *Theory Culture & Society* 25(4):139–58.

Blencowe, C. (2010) 'Foucault's and Arendt's Positive Account of Biopolitics: A Critique of Agamben' *History of the Human Sciences* 23(5):113–30.

Blencowe, C. (2011) 'Biology, Contingency and the Problem of Racism in Feminist Discourse' *Theory Culture & Society* 28(3):3–27.

Booth, R., Lewis, P., & Taylor, M. (2009) 'English Defence League: chaotic alliance stir up trouble on streets', *The Guardian*. London.

Bourdieu, P. (1993) *The Field of Cultural Production*. Cambridge: Polity.

Bourdieu, P., & Passeron, J.-C.(1977) *Reproduction in Education, Society and Culture*. London: Sage Publications.

Braun, K. (2007) 'Biopolitics and Temporality in Arendt and Foucault', *Time & Society* 16(1):5–23.

Burchell, G. (1996) 'Liberal government and techniques of the self' in Barry, Osborne & Rose (eds.) *Foucault and Political Reason: Liberalism, Neo-Liberalism and Rationalities of Government*. London: UCL Press.

Burchill, G., Gordon, C., & Miller, P. (1991) *The Foucault Effect: Studies in Governmentality*. Chicago: Chicago University Press.

Burchill, J. (2005) 'Yeah but no but, why I'm proud to be a Chav', *The Times*, London.

Butler, J. (1988) 'Performative Acts and Gender Construction: An Essay in Phenomenology and Feminist Theory', *Theatre Journal* 40(4):519–38.

Butler, J. (1999) *Gender Trouble: Feminism and the Subversion of Identity*. 10th Anniversary Edition. New York; London: Routledge.

Butler, J. (2004) 'Variations on Sex & Gender: Beauvoir, Wittig, Foucault', in *The Judith Butler Reader*. Malden, MA.; Oxford: Blackwell.

Butler, J. (2009) *Frames of War: When is Life Grievable?* London: Verso.

Buzl, M. et al. (2007) *Anti-Semitism & Islamaphobia: Hatreds Old & New in Europe*. Chicago: Prickly Paradigm Press.

Caygill, H. (1998) *Walter Benjamin: The Colour of Experience*. London: Routledge.

Cooper, M. (2008) *Life as Surplus: Biotechnology and Capitalism in the Neoliberal Era*. Seattle, Wash.: University of Washington Press.

Dean, M. (1996) 'Foucault, Government and the Enfolding of Authority' in Barry, Osborne & Rose (eds.) *Foucault and Political Reason: Liberalism, Neo-Liberalism and Rationalities of Government*. London: UCL Press.

Deleuze, G. (1988) *Foucault*. London: Athlone.

Deleuze, G. (1991) *Bergsonism*. New York: Zone.

Deleuze, G. (1995) *Negotiations*. New York: Colombia.

Deleuze, G. & Guattari F. (1988) *A Thousand Plateaus*. London: Athlone.

Deleuze, G. & Guattari F. (2004) *Anti-Oedipus*. London: Continuum.

Dillon, M. (2005) 'Cared to Death: The Biopoliticised Time of Your Life', *Foucault Studies* 2:37–46.

Dillon, M & Reid, J. (2009) *The Liberal Way of War: The Martial Face of Global Biopolitics*. Oxford: Routledge.

Dolan, F. (2005) 'The Paradoxical Liberty of Bio-Power: Hannah Arendt and Michel Foucault on Modern Politics' *Philosophy & Social Criticism* 31(3):369–380.

Dreyfus, H. & Rabinow, P. (1983) *Beyond Structuralism & Hermeneutic*, Chicago: University of Chicago Press.

Duffield, M. (2007) *Development, Security and Unending War: Governing the World of Peoples*. Cambridge: Polity.

Duffield, M. R., & Hewitt, V. M. (2009) *Empire, Development & Colonialism: The Past in the Present*. Woodbridge, Suffolk; Rochester, NY: James Currey.

EDL. (2010) English Defence League. www.englishdefenceleaguw.org last visited June 2010.

Esposito, R. (2008) *Biòs: Biopolitics and Philosophy*. Minneapolis, MN: University of Minnesota Press.

Foucault, M. (1970) *The Order of Things: An Archaeology of the Human Sciences*. London: Tavistock Publications.

Foucault, M. (1973) *The Birth of the Clinic: an Archaeology of Medical Perception*. London: Routledge.

Foucault, M. (1977) *Discipline and Punish: the Birth of the Prison*. London: Allen Lane.

Foucault, M. (1978) *The History of Sexuality Volume 1 – the Will to Knowledge*. Harmondsworth: Penguin.

Foucault, M. (1980) *Power/Knowledge*, C. Gordon (ed.). Brighton: Harvester.

Foucault, M. (1985a) *The History of Sexuality Volume 2 – The Use of Pleasure*. London: Penguin.

Foucault, M. (1985b) 'Croître et multiplier – *Le Monde* 1970 (sur F. Jacob *La Logique du Viviant: une histoire de l'hérédité*, Paris Gallimard, 1970)' in D. Defert & F. Ewald (eds.), *Dits et ecrits: II 1970–1975*. Paris: Gallimard.

Foucault, M. (1988) *The History of Sexuality Volume 3 – The Care of the Self*. New York: Vintage.

Foucault, M. (1991) *Remarks on Marx: Conversations with Duccio Trombadori*. New York: Columbia University Press.

Foucault, M. (1996) *Foucault Live: (Interviews, 1961–1984)*. New York: Semiotext(e).

Foucault, M. (2000a) 'Truth and Power' in J. D. Faubion (ed.), *Power: Essential Works of Foucault Volume 3*. London: Penguin.

Foucault, M. (2000b) 'Life: Experience and Science' in J. D. Faubion (ed.), *Aesthetics: The Essential Works of Michel Foucault, Volume. 2*. London: Penguin.

Foucault, M. (2000c) 'Interview with Michel Foucault' in J. D. Faubion (ed.), *Power: Essential Works of Foucault Volume 3*. London: Penguin. n.b. – this is an extensive extract of *Remarks on Marx* (Foucault,1991).

Foucault, M. (2000d) 'Truth and Juridicial Forms' in J. D. Faubion (ed.), *Power: Essential Works of Foucault Volume 3*. London: Penguin.

Foucault, M. (2000e) 'Questions of Method' in J. D. Faubion (ed.), *Power: Essential Works of Foucault Volume 3*. London: Penguin.

Foucault, M. (2000f) 'Structuralism and Post-structuralism' in J. D. Faubion (ed.), *Aesthetics: The Essential Works of Michel Foucault, Volume 2*. London: Penguin.

Foucault, M. (2000g) 'On the Archaeology of the Sciences: Response to the Epistemology Circle' in J. D. Faubion (ed.), *Aesthetics: The Essential Works of Michel Foucault, Volume 2*. London: Penguin.

Foucault, M. (2000h) 'The Subject and Power' in J. D. Faubion (ed.), *Power: Essential Works of Foucault Volume 3*. London: Penguin.

Foucault, M. (2000i) ' "*Omnes et Singulatium*": Toward a Critique of Political Reason' in J. D. Faubion (ed.), *Power: Essential Works of Foucault Volume 3*. London: Penguin.

Foucault, M. (2000j) 'About the Concept of the "Dangerous Individual" in Nineteenth-century Legal Psychiatry' in J. D. Faubion (ed.), *Power: Essential Works of Foucault Volume 3*. London: Penguin.

Foucault, M. (2000k) 'Space, Knowlegde, and Power' in J. D. Faubion (ed.), *Power: Essential Works of Foucault Volume 3*. London: Penguin.

Foucault, M. (2000l) P. Rabinow (ed.) *Ethics: The Essential Works of Michel Foucault, Volume. 1*. London: Penguin.

Foucault, M. (2003a) *Abnormal: Lectures at the Collège de France 1974–1975*. London: Verso.

Foucault, M. (2003b) *Society Must Be Defended: Lectures at the Collège de France, 1975–76*. London: Allen Lane.

Foucault, M. (2007) *Security, Territory, Population: Lectures at the Collège de France 1977–1978*. Basingstoke: Palgrave Macmillan.

Foucault, M. (2008a) *The Birth of Biopolitics: Lectures at the Collège de France 1978–1979*. Basingstoke: Palgrave Macmillan.

Foucault, M. (2008b) *Le Gouvernement de Soi et des Autres – Cours au Collège de France 1982–1983*. Paris: Gallimard: Seuil.

Franks, A. (2005) *Margaret Sanger's Eugenic Legacy: The Control of Female Fertility*. Jefferson, NC: McFarland.

Fraser, D. (2003) *The Evolution of the British Welfare State: A History of Social Policy since the Industrial Revolution*. Houndmills, Basingstoke, Hampshire; New York: Palgrave Macmillan.

Fraser, M. (2002) 'What is the matter of feminist criticism?', *Economy and Society*, 31(4):606–25.

Fraser, M., Kember, S. & Lury, C. (2005) 'Inventive Life: Approaches to the New Vitalism', special edition of *Theory, Culture & Society* 22(1).

Fraser, N. (1981) 'Foucault on Modern Power: Empirical Insights and Normative Confusions', *Praxis International* (3):272–87.

Fuss, D. (1989) *Essentially Speaking: Feminism, Nature & Difference*. New York: Routledge.

Gatens, M. (1996) *Imaginary Bodies: Ethics, Power and Corporeality*. London: Routledge.

Gatens, M., & Lloyd, G. (1999) *Collective Imaginings: Spinoza, Past and Present*. London: Routledge.

Gilroy, P. (1987) *There Ain't No Black in the Union Jack: The Cultural Politics of Race & Nation*. London: Hutchinson.

Gordon, C. (1987) 'The Soul of the Citizen: Max Weber and Michel Foucault on Rationality and Government', in S. Lash & S. Whimster (eds.), *Max Weber, rationality and modernity*. London; Allen & Unwin.

Gordon, L. (1990) in L. Gordon (ed.) *Women, the State and Welfare*. Madison, WI: University of Wisconsin Press.

Grimes, S. (1998) 'From Population Control to "Reproductive Rights": Ideological Influences in Population Policy', *Third World Quarterly* 19(3):375–93.

Grosz, E. A. (1994) *Volatile Bodies: Toward a Corporeal Feminism*. Bloomington, IN: Indiana University Press.

Grosz, E. A. (1995) *Space, Time, and Perversion: Essays on the Politics of Bodies*. New York: Routledge.

Grosz, E. A. (2002) 'A politics of imperceptibility', *Philosophy and Social Criticism* 28(4):463–72.

Grosz, E. A. (2004) *The Nick of Time: Politics, Evolution, & the Untimely*. Durham, NC: Duke University Press.

Gutting G. (2001) *French Philosophy in the Twentieth Century*. Cambridge: University of Cambridge Press.

Habermas, J. (1987) *The Theory of Communicative Action*. Cambridge: Polity.

Hacking, I. (1982) 'Biopower and the Avalanche of Printed Numbers', *Humanities in Society* 5:279–95.

Hall, S. (2000) 'Conclusion: The Multicultural Question' in B. Hesse (ed.) *Un/settled Multiculturalisms: Diasporas, Entanglements, Transruption.*, London: Zed Books.

Hallward, P. (2001) *Absolutely Postcolonial: Writing Between the Singular and Specific*. Manchester: Manchester University Press.

Hallward, P. (2006) *Out of This World: Deleuze and the Philosophy of Creation*, London, Verso.

Haraway, D. J. (1991) *Simians, Cyborgs and Women: the Reinvention of Nature*. London: Free Association.

Hardt, M., & Negri, A. (2000) *Empire*, Cambridge, MA: Harvard University Press.

Hardt, M. & Negri, A. (2005) *Multitude*. London: Penguin.

Heron, B. (2007) *Desire for Development: Whiteness, Gender and the Helping Imperative*. Ontario: Wilfred Laurier University Press.

Horizon. (1995) 'The Human Laboratory', *Horizon*. London, BBC.

Irigaray, L. (2004) *Key Writings*. London: Continuum.

Jacob, F. (1973) *The Logic of Life: The History of Heredity*. New York: Pantheon.

Jackson, S. (1995) 'Gender and Heterosexuality: A Materialist Feminist Analysis' in M. Maynard, J. Purvis & Women's Studies Network (UK) Association. (eds), *(Hetero)Sexual Politics*. London: Taylor & Francis.

Jackson, S., & Rees, A. (2007) 'The Appalling Appeal of Nature: the Popular Influence of Evolutionary Psychology as a Problem for Sociology', *Sociology-the Journal of the British Sociological Association* 41(5):917–30.

Jay, M. (2006) *Songs of Experience: Modern American and European Variations on a Universal Theme*. Berkeley, CA: University of California Press.

Juniper, J & Jose, J. (2008) 'Foucault and Spinoza: Philosophies of Immanence and the Decentred Political Subject', *History of the Human Sciences* 21(2):1–20.

Kistner, U. (1999) 'Georges Cuvier: Founder of Modern Biology (Foucault), or Scientific Racist (Cultural Studies)', *Configerations* 7(2):175–190.

Kothari, U. (2006) 'An Agenda for Thinking about "Race" in Development', *Progress in Development Studies* 6(1):9–23.

Lawson, V. (2007) *Making Development Geography*. Oxford: Oxford University Press.

Lash, S. (2007) 'Power After Hegemony: Cultural Studies in Mutation?', *Theory, Culture & Society* 24(3):55–78.

Lazzarato, M. (2002) 'From Biopower to Biopolitics', *The Warwick Journal of Philosophy*, 13: 112–25.

Lazzarato, M. (2009) 'Neoliberalism in Action Inequality, Insecurity and the Reconstitution of the Social', *Theory Culture & Society* 26(6):109–33.

Levitas, R. (2003) 'On Dialectical Utopianism', *History of the Human Sciences* 16(1):137–50.

Lewis, J. (2000) 'Health and Health Care in the Progressive Era' in R. Cooter & J. Pickstone (eds), *Companion to Medicine in the Twentieth Century*. New York: Routledge.

Lorber, J. (1993) 'Believing Is Seeing – Biology as Ideology', *Gender & Society* 7(4):568–81.

Macherey, P. (1992) 'Towards a Natural History of Norms', in T. J. Armstrong (trans; ed), *Michel Foucault, Philosopher: Essays Translated from the French and German*. New York: Harvester Wheatshaft.

Mann, S., & Huffman, D. (2005) 'The Decentring of Second Wave Feminism and the Rise of the Third Wave', *Science & Society* 69(1):56–91.

Massumi, B. (2005) 'The Future Birth of the Affective Fact', *Genealogies of Biopolitics* (conference papers): Radical Empiricism: radicalempiricism.org.

Massumi, B. (2009) 'National Enterprise Emergency Steps Toward an Ecology of Powers', *Theory Culture & Society* 26(6):153–85.

McGrane, B. (1989) *Beyond Anthropology: Society and the Other*. New York: Columbia University Press.

McNay, L. (1994) *Foucault: A Critical Introduction*. Oxford: Polity Press.

Meikle, J. (2009) 'Past lives: 1911 census goes online: records of 27m people lift lid on celebrity ancestors and women's fight for vote', *The Guardian*, 13 January 2009, London.

Merquior, J. G. (1985) *Foucault*. Los Angeles: University of California Press.

Mohanty, C. T. (1988) 'Under Western Eyes, Feminist Scholarship and Colonial Discourses', *Feminist Review* 30:61–88.

Mohanty, C. T. (2003) '"Under Western Eyes" revisited: Feminist solidarity through anticapitalist struggles', *Signs* 28(2):499–535.

Mottier, V. & Gerodetti, N. (2007) 'Eugenics and Social Demoncracy: or How the European Left tried to eliminate the 'Weeds' from its National Gardens' *New Formations* 20: 35–49.

Narayan, U. (2000) 'Essence of Culture and a Sense of History: A Feminist Critique of Cultural Essentialism' in U. Narayan & S. Harding (eds), *Decentreing the Center: philosophy for a multicultural, postcolonial and feminist world*. Bloomington, IN: University of Indiana Press.

Ojakangas, M. (2005) 'Impossible Dialogue on Bio-power: Agamben and Foucault' *Foucault Studies* 2:5–28.

Osborne, T. (1996) 'Security and Vitality: Drains, Liberalism and Power in the Nineteenth Century' in Barry, Osborne & Rose (eds.) *Foucault and Political Reason: Liberalism, Neo-Liberalism and Rationalities of Government*. London: UCL Press.

Osborne, T. (2003a) 'What is a Problem?' *History of the Human Sciences* 16(4):1–17.

Osborne, T. (2003b) 'Against 'creativity': a philistine rant' *Economy & Society* 32(4):507–25.

Osborne, T. (2008) *The Structure of Modern Cultural Theory*. Manchester: Manchester University Press.

Outram, D. (1986) 'Uncertain legislator: Georges Cuvier's laws of nature in their intellectual context' *Journal of the History of Biology* 19(3):323–68.

Owen, D. (2002) 'Criticism and captivity: On genealogy and critical theory', *European Journal of Philosophy* 10(2):216–30.

Parisi, L. (2004) *Abstract Sex: Philosophy, Bio-technology and the Mutations of Desire*. London; New York: Continuum.

Parisi, L. (2009) 'An Archigenesis of Experience', *Australian Feminist Studies* 24(59):31–51.

Petchesky, R. P. (2003) *Global Prescriptions: Gendering Health and Human Rights*. New York: Zed Books.

Phillips, A. (n.d.) 'Feminist Politics: Facing the Future' *Women IT Gender Archive* www.women.it/cyberarchive/files/phillips.htm, last visited December 2009.

Potter, M. (2007) 'Blacks to blame for violence: Blair', *The Toronto Star*. Toronto.

Power, N. (2009) *One Dimensional Woman*. London: Zero Books.

Rabinow, P. (1999) 'Artificiality and Enlightenment: From Sociobiology to Biosociality', in C. Samson (ed), *Health Studies: A Critical and Cross-Cultural Reader*. Oxford: Blackwell Publishers.

Rheinberger, H. J. (2008) 'What Happened to Molecular Biology?', *BioSocieties* 3(3):303–10.

Richardson, A. (1999) 'The Eugenization of Love: Sarah Grand and the Morality of genealogy', *Victorian Studies* 42(2):227–55.

Richardson, A. (2000) 'Biology and Feminism', *Critical Quarterly* 42(3):35–63.

Robinson, K. (2003) 'The Passion and the Pleasure – Foucault's Art of not Being Oneself', *Theory Culture & Society* 20(2):119–44.

Rose, N. (1989) *Governing the Soul: The Shaping of the Private Self*. London: Free Association Books.

Rose, N. (1999) *Powers of Freedom*, Cambridge: Cambridge University Press.

Rose, N. (2001) 'The politics of life itself', *Theory Culture & Society* 18(6):1–30

Rose, N. (2007) *Politics of Life Itself: Biomedicine, Power and Subjectivity in the Twenty-First Century*. Princeton, NJ; Oxford, Princeton University Press.

Rose, N. & Rabinow, P. (2003) 'Thoughts on the Concept of Biopower Today', www.molsci.org/research/publications_pdf/Rose_Rabinow_Biopower_Today. pdf, last visited 18th April 2008.

Rose, N., O'Malley, P., & Valverde, M. (2006) 'Governmentality', *Annual Review of Law and Social Science* 2:83–104.

Rosenberg, R. (1975) 'In Search of Woman's Nature, 1850–1920', *Feminist Studies* 3(1/2):141–154.

Salih, S. & Butler, J. (2003) *The Judith Butler Reader*. Malden, MA: Blackwell.

Scharnhorst, G. (2000) 'Historicising Gilman: A Bibliographer's View' in C. Golden & J. S. Zangrando (eds.), *The Mixed Legacy of Charlotte Perkins Gilman*. London: Associated University Presses.

Schroeder, W. R. (2005) *Continental Philosophy: a Critical Approach*. Malden, MA, Oxford: Blackwell.

Scott, D. (2003) 'Culture in Political Theory', *Political Theory* 31(1):92–115.

Scott, J. W. (1991) 'The Evidence of Experience', *Critical Inquiry* 17(4):773–97.

Senellart, M. (2003) 'Course Context' in Foucault, *Security, Territory, Population: Lectures at the Collège de France 1977–197*, Basingstoke: Palgrave Macmillan.

Shu, J. (2006) 'Women in Fascist Biopolitics: The Case of Olive Hawkes', *Womens Studies* 35:265–84.

Simmel, G. (1971) 'The Transcendent Character of Life' in Levine, T. ed. *Georg Simmel on Individuality and Social Forms*. Chicago: University of Chicago Press.

Skinner, D. (2006) 'Racialized Futures: Biologism and the Changing Politics of Identity', *Social Studies of Science* 36(3):459–88.

Skinner, D. (2007) 'Groundhog Day? The Strange Case of Sociology, Race and 'Science'', *Sociology – The Journal of the British Sociological Association* 41(5):931–43.

Stocking, G. W. (1968) *Race, Culture, and Evolution; Essays in the History of Anthropology*. New York: Free Press.

Stoler, A. L. (1995) *Race and the Education of Desire: Foucault's History of Sexuality and the Colonial Order of Things*, Durham, NC: Duke University Press.

Taylor, M. (2010) 'English Defence League: Inside the Violent World of Britain's New Far-Right', *The Guardian Online*. London.

Taylor Allen, A. (2000) 'Feminism and Eugenics in Germany and Britain, 1900–1940: A Comparative Perspective', *German Studies Review* 23(3):477–505.

Taylor Allen, A. (1993) 'Feminism, Venereal-Diseases, and the State in Germany, 1890–1918', *Journal of the History of Sexualit*, 4(1):27–50.

Thrift, N. (2001) 'Summoning Life', in P. J. Cloke, P. Crang & M. Goodwin (eds.), *Envisioning Human Geographies*. London: Arnold.

Thrift, N. (2007) 'Overcome by Space: Reworking Foucault', in J. W. Crampton & S. Elden (eds), *Space, Knowledge and Power: Foucault and Geography*. Aldershot: Ashgate.

UCMP. (n.d.) 'Georges Cuvier (1769–1832)', *University of California Museum of Paleontology*, www.ucmp.berkeley.edu/history/cuvier.html, last visited January 2009.

Valverde, M. (2007) 'Genealogies of European States: Foucauldian Reflections', *Economy and Society* 36(1):159–78.

Walkowitz, J. R. (1980) *Prostitution and Victorian Society: Women, Class, and the State*. Cambridge; New York: Cambridge University Press.

Walkowitz, J. R. (1992) *City of Dreadful Delight: Narratives of Sexual Danger in Late-Victorian London*. Chicago: University of Chicago Press.

Warwick, D. P. (1982) *Bitter Pills: Population Policies and Their Implementation in Eight Developing Countries*. Cambridge: Cambridge University Press.

Weikart, R. (2004) *From Darwin to Hitler*. London: Palgrave Macmillan.

Wittig, M. (1993) 'One Is Not Born a Woman', in H. Abelove, M. A. Barale & D. M. Halperin (eds), *The Lesbian and Gay Studies Reader*. New York: Routledge.

Wittig, M. (1971) *The Guérillères*. London: Picador.

Wittig, M., & Zeig, S. (1979) *Lesbian Peoples: Material for a Dictionary*. New York: Avon.

Ziegler, M. (2008) 'Eugenic Feminism: Mental Hygiene, The Women's Movement, and the Campaign for Eugenic Legal Reform, 1900–1935', *Harvard Journal of Law and Gender* 31:211–35.

Index

Page numbers in **bold** denote definition of term